30 BeagleBone Black Projects for the Evil Genius™

Evil Genius Series

30 BeagleBone Black Projects for the Evil Genius™

Christopher Rush

New York   Chicago   San Francisco   Athens   London   Madrid
Mexico City   Milan   New Delhi   Singapore   Sydney   Toronto

Library of Congress Cataloging-in-Publication Data

Rush, Christopher.
  30 BeagleBone Black projects for the evil genius / Christopher Rush.
    pages cm
  Includes index.
  ISBN 978-0-07-183928-0 (paperback)
  1. Electronic apparatus and appliances—Design and construction—Amateurs' manuals.
  2. Programmable controllers—Amateurs' manuals. 3. BeagleBone Black (Computer)—Amateurs'
  manuals. I. Title. II. Title: Thirty BeagleBone Black projects for the evil genius.
  TK9965.R87   2014
  621.39—dc23

                                                                                    2014029158

McGraw-Hill Education books are available at special quantity discounts to use as premiums and sales promotions, or for use in corporate training programs. To contact a representative, please visit the Contact Us pages at www.mhprofessional.com.

30 BeagleBone Black Projects for the Evil Genius™

1 2 3 4 5 6 7 8 9 0   QVS QVS   1 0 9 8 7 6 5 4

ISBN   978-0-07-183928-0
MHID   0-07-183928-3

Sponsoring Editor Roger Stewart	**Proofreader** Lisa McCoy
Editorial Supervisor Patty Mon	**Indexer** James Minkin
Project Manager Patricia Wallenburg, TypeWriting	**Production Supervisor** Jean Bodeaux
Acquisitions Coordinator Amy Stonebraker	**Composition** TypeWriting
Copy Editor Bart Reed	**Art Director, Cover** Jeff Weeks

I would like to dedicate this book to my father, Peter Rush,
from whom I inherited the love to make things.

About the Author

Christopher Rush (Preston, UK) has a degree in computer science and has spent the last 10 years working for CPC, an electronics distribution company. Christopher is a full-time product author and has an extensive knowledge of the electronics hobbyist market. Christopher also runs a makerspace blog, providing tutorials and user guides for popular development boards and accessories, including Raspberry Pi, Arduino, BeagleBone, and others.

Contents

Acknowledgments

I WOULD LIKE TO THANK my family for all their support, and my team at CPC for all their encouragement. Also, I would like to say a special thanks to my close friend Dr. Mark Lochrie, who is always very supportive and inspiring, as well as to my partner Jennifer Wozniak for her patience and motivation while writing this book.

I would like to thank Roger Stewart and the team at McGraw-Hill Education, who have been very patient and supportive and an absolute pleasure to work with. And, finally, thanks to Simon Monk—all this would not have been possible without you.

Introduction

THE BEAGLEBONE BLACK PROVIDES the Evil Genius with a low-cost, easy-to-use solution for creating some Evil projects. The board allows us to create a wide range of projects that can be built and controlled by a single embedded solution. Before too long, you will have captured your enemy and used your sinister Evil techniques to interrogate them. They will be at your mercy—mwahahahaha!

This book will show you how to connect your BeagleBone Black to your computer and how to program it using the free software provided. It will also teach you how to connect a wide range of common electronic components, including LEDs, motors, a USB missile launcher, and much more.

This book includes full schematic and breadboard layout diagrams, where appropriate, for most of the projects. Also, most of the projects can be constructed without the need of any soldering equipment or tools. However, the deeper you dive into the more Evil projects, the more complex the creations become, and instructions are clearly provided for soldering components to prototyping boards.

What Is BeagleBone Black All About?

The BeagleBone Black is a small microcontroller computer board specifically designed for developers and hobbyists, supplied with a mini-USB port that plugs directly into your computer. It also features a number of sockets and connectors that can be hooked up to a number of electrical devices and components such as LEDs, motors, sensors, speakers, and many more. The BeagleBone Black can be powered either through the mini-USB port or directly from a 5V power supply unit. When connected to a computer, it can then be controlled and programmed through the computer, but also disconnected for use as a true embedded system.

At this current time, Evil Geniuses can easily obtain a BeagleBone Black from their favorite online electronic store for around $45; this includes a BeagleBone Black board and a mini-USB cable for connectivity.

BeagleBone Black

The BeagleBone Black isn't just an open source embedded platform; it is a whole ecosystem of electronics and interface boards encompassing the enthusiasm from the open source community. Never underestimate the power of community—there is a large community of engineers, hobbyists, makers, and Evil Geniuses out there, most of which can be found on Beagleboard.org. These makers create and share their ideas and projects online with engaged enthusiasts. The following table lists the key hardware features of the BeagleBone Black.

	Feature
Processor	Sitara AM3358BZCZ100 1 GHz, 2,000 MIPS
Graphics	SGX530 3D, 20M Polygons/S
SDRAM Memory	512MB DDR3L 800 MHz
Onboard Flash	2GB, 8-bit Embedded MMC
PMIC	TPS65217C PMIC regulator and one additional LDO
Debug Support	Optional onboard 20-pin CTI JTAG, serial header
Power Source	Mini USB or 5VDC jack
PCB	3.4" × 2.1", six layers
Indicators	1: Power 2: Ethernet 4: User-controllable LEDs
USB 2.0 Client Port	Via mini USB
USB 2.0 Host Port	Type A socket 500mA
Serial Port	Six-pin 3.3V TTL header
Ethernet	10/100 RJ-45
SD/MMC Connector	Micro-SD
User Input	Reset, Boot, Power button
Video Output	HDMI 1280×1024
Audio	Via HDMI interface
Expansion Connectors	Power 5V, 3.3V, VDD_ADC (1.8V) 3.3V I/O on all signals 65 GPIO, I2C, SPI, 7 AIN (1.8V max), four timers, four serial ports
Weight	40g (approx.)
Power	210–460mA @ 5V

This book uses Revision A5C with the latest version of the Angstrom operating system. Full instructions on obtaining the latest version can be found in Chapter 1.

A full list of the BeagleBone Black's GPIO pins can be found in Appendix B. This is very useful when you are deciding which electronic devices to attach to the BeagleBone.

Capes

Capes are expansion boards that connect directly on top of the BeagleBone Black. They are designed by developers and advanced makers as an easy solution for creating your projects. Currently, over 50 capes are available for both the BeagleBone and BeagleBone Black, which can be found at http://elinux.org/Beagleboard:BeagleBone_Capes. Note that some capes are not fully compatible with the BeagleBone Black. The majority of the capes have been produced by the user community; you can submit your cape ideas to the Beagleboard website. Some capes include the following types:

- LCD screen
- Battery cape
- GPS/GPRS cape
- Audio cape
- Prototyping cape
- HD camera cape

Revision C

As of the writing this book, and the first birthday of the BeagleBone Black, a new revision of the BeagleBone Black is available. It ships standard with Debian OS instead of Angstrom. It also comes supplied with 4GB of eMMC flash memory, as requested by the BeagleBone community. This helps prevent the wearing out of flash memory over time. All other features of the board remain the same; however, there is a $10 price increase to cope with the ever-fluctuating costs of materials to manufacture the boards. All the projects in this book are compatible with all revisions of the BeagleBone Black, as well as both Debian and Angstrom. Note that both commands for installing software are stated, where possible.

The Projects

The projects in this book offer a wide range of applications, beginning with some basic examples, such as connecting an LED and getting it to flash on and off. At the same time, you get experience using the software and the BeagleBone Black's GPIO pins. Most of the projects are programmed using the JavaScript BoneScript library, but some projects later in the book use Python and the Adafruit BBIO library. I have tried to accommodate for using both programming languages on the BeagleBone Black.

The chapters move from using LEDs to connecting a range of sensors for detecting temperature and measuring light levels, for example. Although these are still basic projects, they are very useful in everyday applications. We also use a number of already-made modules, such as an LED matrix display and a GPS sensor for obtaining current location statistics.

Gradually building on the previous chapters, we move on to Chapter 4, where you learn how to control the motors and servos used in robotics. You can control them through the web interface or via a standard wireless keyboard. We also create a watering system using a motor pump that detects the moisture level in the soil and then turns on the motor when the plant needs watering—simple, yet satisfyingly useful for those with a green thumb.

The final "projects" chapter, Chapter 7, contains a mixture of some really Evil projects, including a lie detector test, a webcam doorbell security system, a motion detection alert using the Twitter API Tweepy, and a dog-barking deterrent for unwanted guests.

You can build most of the projects in the book without the need for soldering any components. Instead, we use a *breadboard,* which is a plastic block that contains lots of small holes, arranged in rows. These rows are connected using metal springs behind them and are used to create our circuits. More information about breadboards can be found in Chapter 8. If you decide that you want to make your project a more permanent fixture, then refer to Chapter 2, which shows you how to solder your project to a PCB prototyping cape.

The components in this book are all sourced from a number of suppliers in both the United States and the United Kingdom. Appendix A features a full list of parts, part numbers, and suppliers from which you can purchase the parts.

Software

All the software written for this book is freely available for download and is all open source, which means you can reuse the code for your own projects and distribute it as you like. Most of the projects written for this book use the BoneScript programming language and use the .js file extension. Some projects later in the book use the Python language and use the .py file extension.

The software is available to download from https://github.com/ChristopherRush/BB-Evil-Genius. To download the software using github, open a Terminal window from within Cloud9 on Debian and type the following command to download the github repository:

```
git clone git://github.com/
ChristopherRush/BB-Evil-Genius.git
```

If you are using an older version of the BeagleBone Black with Angstrom, you can SSH into the BeagleBone Black and type in the same command to begin the download.

The software code is well annotated, with useful hints and a breakdown of what parts of the code do what.

And I Present to You...

Without further ado, I present to you *30 BeagleBone Black Projects for the Evil Genius.* I know some of you are impatient Evil Geniuses and want to get straight to the good stuff. However, Chapter 1 gives you a step-by-step guide on how to set up your BeagleBone Black with ease and start programming your first project using the Cloud9 IDE program. I highly recommend you read this chapter if you are new to the BeagleBone Black.

Starting in Chapter 2, we begin to build on what you learn in the first chapter, gradually increasing the complexity of the projects and hopefully allowing you to go on and design your own Evil Genius projects. You may want to skip certain chapters and instead select a particular project to get started with—if so, just pick a project and get going. However, if you get stuck, it may be worth starting from the beginning in Chapter 1 and slowly working your way through each project.

Getting Started

LIKE MANY MAKERS, I had started out making electronics using the Arduino and BASIC stamp microcontroller boards. Since then, I have transitioned onto using many other boards such as Freescale 8/16bit, MSP430 chipset boards, Digilent Chipkit, and Raspberry Pi. Although these boards are great, I always found myself wanting more—there was always something lacking in their features or they were not cost effective for my projects.

I heard about the Beagleboard and BeagleBone a while back, but it didn't grab my attention at the time, with no video output and a heavy price tag of $100–$150. When I read a press release from CircuitCo about a new BeagleBone Black, I thought it was too good to be true: the specification was incredible, with an Arm Cortex A8 1GHz CPU with 65 possible GPIO pins, HDMI video/audio output, and 512MB of DDR3 memory. Surely this would come with higher price tag than that of a small mobile computer. Well, I was wrong. CircuitCo announced that the price would be around $45—an incredible value.

Since receiving my BeagleBone Black, I have been hooked. All my projects use this microcontroller, and the more I learn about this board, the more features I unravel and the more it shows its true full potential as the ultimate microcontroller board. This chapter unravels the unknown mysteries of the BeagleBone Black. It gets you familiar with using the hardware,

software, and setting up your first project using the BoneScript programming language.

I know that having so much flexibility and versatility in a single device can sometimes make things seem hard when you are just getting started. You will soon realize that there isn't a single right way of doing things: there can be many different ways of achieving the same outcome. Hopefully, this chapter will get you heading in the right direction.

Powering Up Your BeagleBone Black

You have many ways to connect to your BeagleBone Black development board—more specifically, you have many ways to access it to start programming using BoneScript. The preferred software tool for programming the BeagleBone Black is the Cloud9 IDE, but we'll cover every option for programming in this chapter.

Connecting Using USB

The BeagleBone Black comes with software/drivers already embedded in its operating system (OS) and includes documentation that will help you get connected to your computer. A great new hardware feature for the BeagleBone Black is that it comes with 2GB eMMC Flash memory with the Angstrom OS preinstalled, so there is no need to download and flash a micro-SD card.

Go ahead and plug the mini universal serial bus (USB) cable supplied with your BeagleBone Black (BBB) board into your BeagleBone Black and the other end into your computer, as shown in Figure 1-1.

This will power the BeagleBone Black board and also provide a development interface in which you can program. The BeagleBone Black will boot Angstrom OS from the onboard 2GB embedded multimedia controller (eMMC), and the new revision C board will boot Debian OS. Once the BeagleBone Black is plugged in, the power LED (light-emitting diode) will be lit up and the adjacent bank of four LEDs will be flashing, as shown in Figure 1-2. The board is now alive!

The four LEDs are configured as follows:

- USR0 is configured at bootup to blink in a heartbeat pattern.

- USR1 is configured at bootup to light during micro-SD card access.

- USR2 is configured at bootup to light during CPU activity.

- USR3 is configured at bootup to light during eMMC access.

Figure 1-2   BeagleBone Black's bank of four LED indicators

NOTE   Holding down the boot switch when powering your BeagleBone Black will tell the hardware to boot from the micro-SD card instead of the onboard eMMC.

Installing Drivers

Drivers need to be installed on your computer system before you can do anything with the board, so the following instructions show you how to install drivers on the Windows, Mac OS X, and Linux operating systems. Once you have powered up your BeagleBone Black through USB, it will operate as a flash drive, giving you all the necessary files you need to get started, including drivers and documentation, as shown in Figure 1-3.

Open the file directory on the flash drive and double-click start.html to open this document in your default web browser. The web page you see before you is a step-by-step quick start guide on installing all the relevant software. It provides web links for all stages, making it easy to follow. Go to step 2, shown in Figure 1-4, and select the operating system you are currently using; if prompted, click Run and follow the instructions to install the drivers.

Figure 1-1   Connecting a USB cable to BBB

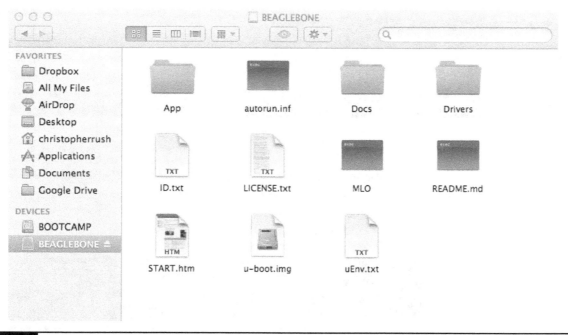

Figure 1-3 The file directory for BBB

Figure 1-4 Selecting operating system drivers

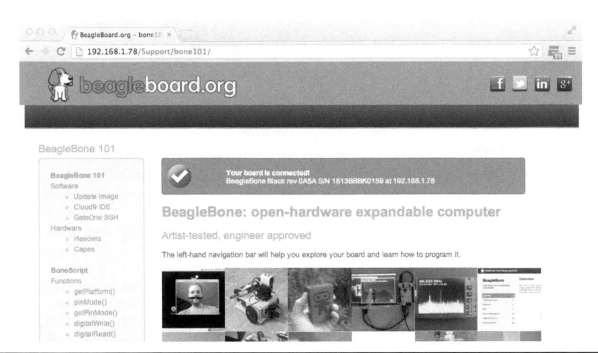

Figure 1-5 The Bone101 web page

> **NOTE** You must install the "FTDI USB to serial/JTAG" drivers if you are connecting through USB.

Once this step is complete, open your web browser (either Firefox or Chrome, not Internet Explorer) and type in the URL http://192.168.7.2/. This will confirm the connection to your BeagleBone Black and load up the Bone101 web page on the device, as shown in Figure 1-5.

Congratulations, you are now one step closer to becoming an Evil Genius.

Connecting to Your BeagleBone Black Through Ethernet

Connecting your BeagleBone Black to your Ethernet network is very convenient when you are connecting using different computers in different locations. This method does require a 5V PSU with a 2.1mm DC jack for powering up the BeagleBone Black. So go ahead and plug

in the Ethernet cable connected to your router/modem and then plug in the DC power jack, as shown in Figure 1-6.

Next, you need to find out what the BeagleBone Black's Internet Protocol (IP) address is so you can access it in your web

Figure 1-6 Ethernet and DC jack connected to BBB

Figure 1-7 BBB IP address

browser. This address is automatically assigned to the BeagleBone Black using the Dynamic Host Control Protocol (DHCP) on your router/modem. The easiest way to find this is to open up your router/modem settings and locate the device under the wired Ethernet settings, like in Figure 1-7.

Once you have located the BeagleBone Black IP address, open your web browser (either Firefox or Chrome, not Internet Explorer). In the URL bar, type the address http://x.x.x.x/, where "x.x.x.x" is your IP address. Once you have done this, press ENTER. You should now be connected to your BeagleBone Black board and see the Bone101 web page, as shown in Figure 1-5.

Congratulations, you are now connected to your BeagleBone Black through an Ethernet connection. The Bone101 page provides a lot of information about the board itself, including some great examples using the BoneScript JavaScript library and some great add-on boards called "capes." Feel free to browse around.

Connecting via SSH Through USB and Ethernet

Secure Shell (SSH) is a cryptographic network protocol for secure data communications between devices, and is more commonly known for remote command execution using a client/server model. Accessing the BeagleBone Black through SSH can be accomplished either by connecting the BeagleBone Black to your computer using the USB cable provided or by connecting the BeagleBone Black to your Ethernet network.

SSH clients come standard on most operating systems (except Microsoft Windows). This is not a problem because a load of software is available that can allow you to connect to your BeagleBone Black through SSH, and this section shows you how to do that.

SSH Through GateOne SSH

Due to the fantastic features on the BeagleBone Black, GateOne SSH comes preinstalled. GateOne SSH is a terminal emulator that you can use through your web browser without installing any additional software on your computer. This is by far the easiest method of connecting via SSH, and the most convenient for connecting using different operating systems, even on mobile devices.

The first step is to open your web browser and navigate to the Bone101 web page (http://192.168.1.78/Support/bone101/). In the navigation menu on the left side, you'll see GateOne SSH under the heading Software. Click the link, which directs you to the GateOne SSH web page shown in Figure 1-8.

Once the web page has loaded, you should see a window that looks similar to a terminal window on your BeagleBone Black. At first you will be prompted to enter connection details

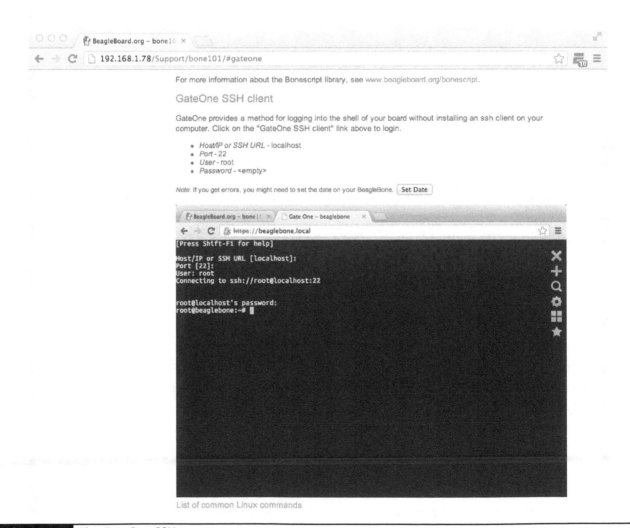

Figure 1-8   The GateOne SSH page

such as IP address, port number, username, and password. If you are connected to the BeagleBone Black through either USB or the local Ethernet network, leave the first line blank when asked for an host/IP address and also when you're prompted to enter a port number, because this instance will be using the default settings to connect. Enter your username and password, and you will be connected to your BeagleBone Black through SSH. You will see "root@ beaglebone:~#" once connected.

 The default username is root, and no password is set.

Connecting via SSH in Windows

To get SSH running in Windows, you need to download an SSH client program. The most popular program to use is PuTTY, and it can be downloaded from www.putty.org. Once it is downloaded and installed, you can go ahead and run the putty.exe program file. You will enter the configuration window upon starting PuTTY (see Figure 1-9). This is where you add your BeagleBone Black settings, which will enable you to connect over SSH.

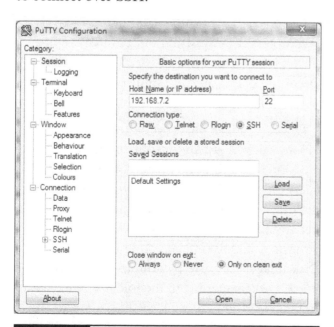

Figure 1-9 The PuTTY Configuration menu

Enter the IP address of your BeagleBone Black in the Host Name box; if you are connected via USB, the IP address will be 192.168.7.2. Alternatively, if you are connected through the local Ethernet network, your BeagleBone Black will have a local IP address that you can obtain through your router/ modem. Next, click Open, and the client will try to establish a connection. A security box will appear; you will be required to enter the default username and password details of the BeagleBone Black. Log in with "root" and no password, after which you can just press ENTER. You are now connected to your BeagleBone Black, as shown in Figure 1-10, and can view the root file system and issue commands.

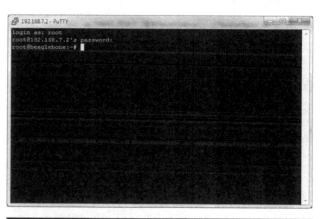

Figure 1-10 The PuTTY window after you have logged in

Connecting via SSH in Mac OS/Linux

If you are a Mac or Linux user, you're in luck: an SSH client comes with the standard installation and can be accessed using the Terminal window (see Figure 1-11). All you have to do is type the following command:

```
ssh 192.168.7.2 —l root
```

```
O O O                    ⌂ christopherrush — ssh — 80×24              ↗
Last login: Sat Jan  4 02:01:31 on console
You have mail.
rushy-pc:~ christopherrush$ ssh 192.168.7.2 -l root
ssh: connect to host 192.168.7.2 port 22: Operation timed out
rushy-pc:~ christopherrush$ ssh 192.168.1.78 -l root
The authenticity of host '192.168.1.78 (192.168.1.78)' can't be established.
RSA key fingerprint is 01:af:6e:a6:d6:cb:0e:52:19:a6:3b:6b:e2:97:3d:fa.
Are you sure you want to continue connecting (yes/no)? yes
Warning: Permanently added '192.168.1.78' (RSA) to the list of known hosts.
root@192.168.1.78's password:
root@beaglebone:~# []
```

Figure 1-11   The SSH login procedure for Mac OS X

The first time you log in, you will receive an authentication warning message. Just type yes to continue with the SSH session and press RETURN (or ENTER) when prompted for a password, given that no default password is set on the BeagleBone Black device.

Installing Software and Running Updates

The BeagleBone Black comes preinstalled with the Angstrom operating system, but sometimes you may want a clean install or you may want to run an update of the current files. I always recommend that you have the latest and most up-to-date operating system and software, so I'll show you how to do both.

Installing the New OS Image

Before we get started installing a new image, a couple of items are required:

- 4GB micro-SD card
- Micro-SD card reader

The BeagleBone Black has 2GB of eMMC storage (4GB on RevC) that can be initialized by a program booted off a micro-SD card.

Download the latest distribution from http://beagleboard.org/latest-images, where you will find the most up-to-date image of the OS.

Depending on your Internet connection speed, the download may take up to 30 minutes.

Writing the Image in Windows

Once the distribution is downloaded, you will notice that the file has an .img.xz extension, which is a compressed sector-by-sector image of the SD card. To decompress this file, you need to download and install 7-Zip, which is available for all operating systems from www.7-zip.org/download.html (see Figure 1-12).

> **NOTE**   Make sure you download BeagleBone Black (eMMC flasher) if you want to replace the current image on the eMMC.

Once 7-Zip is downloaded and installed, decompress the Angstrom image file by right-clicking the image file and selecting 7-Zip | Decompress Image. A decompression in progress is shown in Figure 1-13.

Once the image file is decompressed, you can now download and install an image writer program that will write the image file to the micro-SD card. For Windows computers, you can download Win32 Disk Imager, which is free from SourceForge (http://sourceforge.net/projects/win32diskimager). Launch the Win32 Disk Imager software by double-clicking the Win32DiskImager file in the folder to which you extracted it and then click the file icon next to the Image File field (see Figure 1-14). Navigate

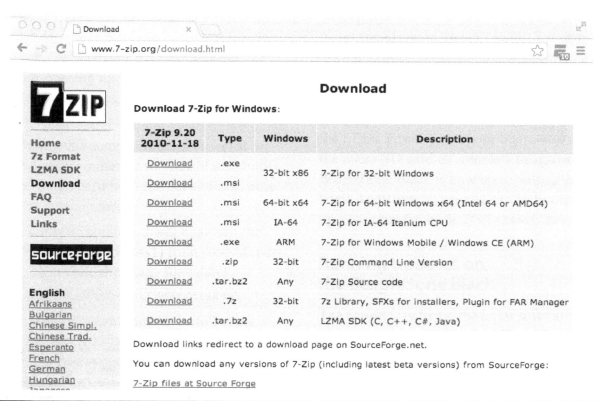

Figure 1-12 The 7-Zip download page

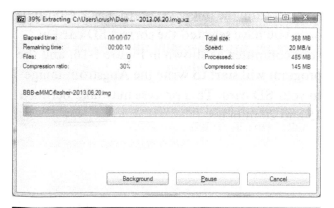

Figure 1-13 7-Zip decompressing the image

Figure 1-14 Win32 Disk Imager

to where you decompressed the Angstrom image and select the image file (see Figure 1-15).

Once this is done, click the Write button and wait for the image file to complete writing to the microSD card.

 **CAUTION** Make sure you select the correct drive letter; otherwise, you risk losing data on other devices!

Writing the Image in Mac OS X

To decompress the image file in Mac OS X, the best program you can use is The Unarchiver, which you can find in the App Store or from http://wakaba.c3.cx/s/apps/unarchiver.html. Once you have downloaded and installed the application, navigate to where the Angstrom image file was downloaded and double-click the

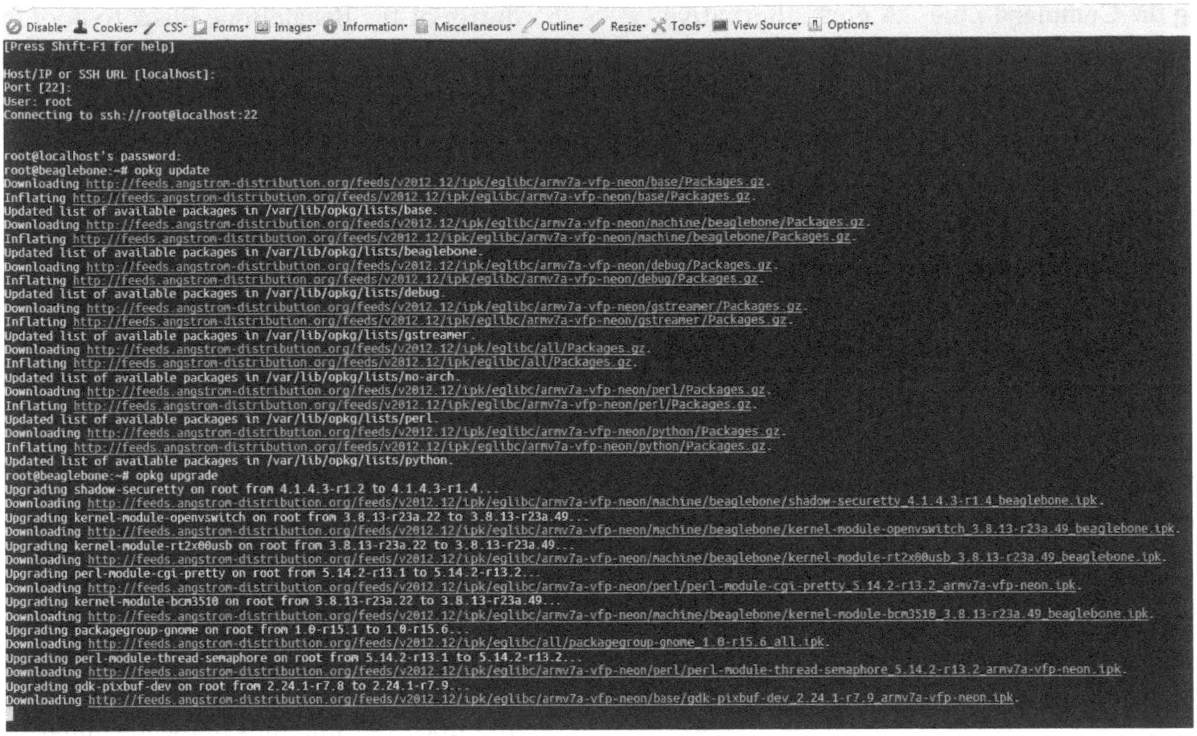

Figure 1-17 Upgrading packages

BeagleBone Black Pinout

Most micro-development platforms provide a range of different inputs and outputs using headers called GPIO (General Purpose Input/ Output) pins. These pins allow you to control things electronically using both hardware and software, and each pin can serve a specific function—either analog or digital. Most microcontrollers come with a pinout diagram with labeled pins stating their function.

The BeagleBone Black comes with two 46-pin dual-row expansion headers, labeled P9 and P8, also known as Expansion A and Expansion B, respectively (see Figure 1-18). Each pin provides 3.3V, unless otherwise stated.

Digital GPIO Pins

The BeagleBone Black comes with 65 GPIO pin headers, which is a huge amount of control at your fingertips. These headers are labeled GPIO_xx, and you can control these pins by switching an output to either "on" or "off" (1 or 0, respectively). You can also detect a digital input signal to sense when a digital input device such as a switch has been activated (on, or 1) or deactivated (off, or 0). These pins are perfect for controlling a vast array of LEDs.

CAUTION Unlike other microcontroller boards, the BeagleBone Black's pins operate using 3.3V. Using anything above this level can permanently damage the board.

Analog Pins

Seven analog pins are available on the BeagleBone Black, labeled AINx. These pins are designed to detect analog signals coming into the BeagleBone Black from such devices as temperature sensors. The BeagleBone Black has a built-in 12-bit ADC function that allows you to convert the in-coming analog signal to a more readable digital value.

P9

DGND	1	2	DGND
VDD_3V3	3	4	VDD_3V3
VDD_5V	5	6	VDD_5V
SYS_5V	7	8	SYS_5V
PWR_BUT	9	10	SYS_RESETn
GPIO_30	11	12	GPIO_60
GPIO_31	13	14	GPIO_40
GPIO_48	15	16	GPIO_51
GPIO_4	17	18	GPIO_5
I2C2_SCL	19	20	I2C2_SDA
GPIO_3	21	22	GPIO_2
GPIO_49	23	24	GPIO_15
GPIO_117	25	26	GPIO_14
GPIO_125	27	28	GPIO_123
GPIO_121	29	30	GPIO_122
GPIO_120	31	32	VDD_ADC
AIN4	33	34	GNDA_ADC
AIN6	35	36	AIN5
AIN2	37	38	AIN3
AIN0	39	40	AIN1
GPIO_20	41	42	GPIO_7
DGND	43	44	DGND
DGND	45	46	DGND

P8

DGND	1	2	DGND
GPIO_38	3	4	GPIO_39
GPIO_34	5	6	GPIO_35
GPIO_66	7	8	GPIO_67
GPIO_69	9	10	GPIO_68
GPIO_45	11	12	GPIO_44
GPIO_23	13	14	GPIO_26
GPIO_47	15	16	GPIO_46
GPIO_27	17	18	GPIO_65
GPIO_22	19	20	GPIO_63
GPIO_62	21	22	GPIO_37
GPIO_36	23	24	GPIO_33
GPIO_32	25	26	GPIO_61
GPIO_86	27	28	GPIO_88
GPIO_87	29	30	GPIO_89
GPIO_10	31	32	GPIO_11
GPIO_9	33	34	GPIO_81
GPIO_8	35	36	GPIO_80
GPIO_78	37	38	GPIO_79
GPIO_76	39	40	GPIO_77
GPIO_74	41	42	GPIO_75
GPIO_72	43	44	GPIO_73
GPIO_70	45	46	GPIO_71

Figure 1-18 BeagleBone Black GPIO pinout

CAUTION Make sure you don't input more than 1.8V to the analog input pins; otherwise, you risk damaging the board.

I2C Pins

The BeagleBone Black contains two I2C pins, labeled I2CX_SCL and I2CX_SDA. The first I2C bus is used for reading EEPROMS on the BeagleBone cape add-on boards and can't be used for other digital I/O operations. However, you can still use it to add other I2C devices. The second I2C bus is available for you to use to configure your devices. The I2C bus allows you to add multiple devices to the BeagleBone Black using the I2C addressing system.

SPI Pins

Two SPI (serial peripheral interface bus) ports are available for use with SPI-compatible devices. SPI is a synchronous data link between devices and operates using a full duplex mode, which enables a faster data transfer. Generally, one device operates as a master and the other a slave in order to synchronize. This also allows you to add multiple slave devices, usually in a daisy chain configuration.

UART Pins

The BeagleBone Black comes with a dedicated header for getting to the UART0 pins and for connecting a debug cable. In addition to the debug header, there are five serial ports in the headers, but only one of them has a single direction.

Project 1
Blinking an Internal LED

Having successfully set up your BeagleBone Black, you are now eagerly anticipating your first project. This project will get you familiar with using the BeagleBone—nothing exciting is going to happen, nor any evildoing! However, it's a start—and we all have to start somewhere. This project requires no additional electrical hardware, thus allowing us to primarily focus on the programming side of things. This also ensures that you have everything set up correctly on your BeagleBone Black board.

So, without further ado, we are going to write a program that will blink the onboard LEDs on the BeagleBone Black. If you have had some previous programming experience, this is going to be our "Hello World" program. We will write this program from scratch so you can get a feel for the structure of using BoneScript in Cloud9 IDE. We will go through the code line by line to give you an explanation of what each function does.

The code for blinking the internal LED is as follows:

```
var b = require('bonescript');
var led = "USR3";

b.pinMode(led, b.OUTPUT);

var state = b.LOW;

b.digitalWrite(led, state);

setInterval(toggle, 1000);

function toggle() {
    if(state == b.LOW) state = b.HIGH;
    else state = b.LOW;
    b.digitalWrite(led, state);
}
```

When writing a program using BoneScript, we need to point our program to the BoneScript library so we can access the GPIO headers and other functions on the BeagleBone Black. Therefore, the first line of our code creates a variable (b) that we can reference in our code to access the BoneScript listed in the parentheses:

```
var b = require('bonescript');
```

The next logical piece of code is to create a new variable so we can refer to the onboard LED labeled USR3. For this instance, we call the variable led, and the string to access the LED is "USR3":

```
var led = "USR3";
```

The GPIO digital pins on the BeagleBone Black can be set as either an input or an output pin, so in our code we need to tell the BeagleBone Black that we want our onboard LED to be an output. To do this, we use a function called pinMode and then select in parentheses the pin we wish to use; in this case, we are using the variable led, and then we set the pin as an output using variable b.OUTPUT:

```
b.pinMode(led, b.OUTPUT);
```

In this program, we create a loop that gathers the state of the LED (either HIGH or LOW) and then alternates between these values to switch the LED on and off. To do this, we need to set another variable, called state, and assign a value to it; in this case, we initially set the LED to "off" (or b.LOW):

```
var state = b.LOW;
```

Now that we have set the variable for the LED, we need to send the command to the digital pin on the BeagleBone Black to set the LED to LOW (off). To do this, we use the function digitalWrite and select the GPIO pin led and choose whether it is HIGH or LOW (that is, we set its "state"):

```
b.digitalWrite(led, state);
```

After the state of the LED has been set, we need to toggle the LED to flash on and off. To do this, we set a time interval of 1,000 milliseconds using the function setInterval. After 1,000 milliseconds, we call the function toggle:

```
setInterval(toggle, 1000);
```

We now have created a function called toggle, and every 1,000 milliseconds, this function will be called. Now we need to alternate the state of the LED between HIGH and LOW.

The easiest way to achieve this is to use an `if` `else` statement to create a different action for different decisions. The statement is either TRUE or FALSE, as shown here:

```
if (condition) TRUE
  {
  code to be executed if condition is
true
  }
else FALSE
  {
  code to be executed if condition is
not true
  }
```

In our function, we check to see if the state is equal to LOW, and if it is (TRUE) we then set

the state to HIGH (that is, turn the LED on). If the state is not equal to LOW, we then trigger the `else` statement (FALSE) and we set the state to LOW (that is, turn the LED off). Once we have set the state (or alternated between ON and OFF, or HIGH and LOW), we then issue the `digitalWrite` function to turn the LED on or off:

```
function toggle() {
    if(state == b.LOW) state = b.HIGH;
    else state = b.LOW;
    b.digitalWrite(led, state);
}
```

Go ahead and run the program by clicking Run in the Cloud9 menu bar, shown in Figure 1-19.

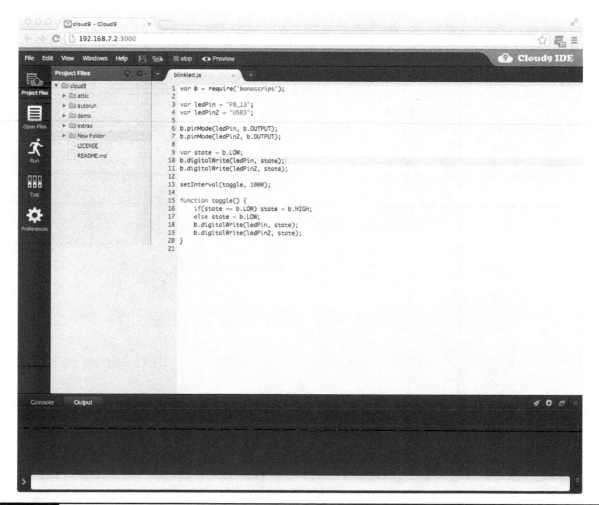

Figure 1-19 The Cloud9 Run program

You should now see the USR3 LED flash on and off every 1,000 milliseconds. If you want to change the interval timing between flashes, you can change the timing in the following line:

```
setInterval(toggle, 1000);
```

Summary

You now have your BeagleBone Black set up and have created your first project. It's not the most exciting Evil project we will conjure, but you now know that everything works and you have learned the basics of how the BeagleBone Black operates. You will move on to more advanced projects as you progress through this book. Along the way, you will gain insight into various ways you can create your own Evil projects.

CHAPTER 2

LED Projects

IN THIS CHAPTER we are going to create some light-emitting diode (LED) projects that will amaze and dazzle you—well, not really, but no doubt you will be impressed. We will keep the hardware very simple so that you have time to familiarize yourself with the BeagleBone Black. This will also give you a better understanding of how the BoneScript works on the programming side of things.

Using LEDs is a great way of interacting and communicating with end users using light. Making your projects stand out can be important and a great way of getting the message across using human–computer interaction (HCI). LEDs are most certainly one of the most used parts in electronic projects. LEDs are cheap and easy to use; they can be used for just about any application and come in a variety of sizes and colors, which is great for creating some very cool effects.

Project 2
Blink an External LED

In this project, we are going to build on the first LED project we created in Chapter 1 by connecting an LED to our BeagleBone Black using the GPIO pins to create a simple electrical circuit that will flash an LED on and off. This will get you familiar with using the GPIO pins on the BeagleBone Black and using the GPIO

pin-naming convention. It will also give you an understanding of how LEDs work.

Hardware

Table 2-1 lists the components and equipment required for Project 2. The hardware is very simple; you can use any standard 5/10mm LED in any color for this project. The breadboard is a half-size 2.54mm pitch between pins and also has power rails down either side for ease of use. See Chapter 8 for more detail about using breadboards.

TABLE 2-1	Components and Equipment for Project 2	
Schematic Reference	Description	Appendix A
	BeagleBone Black	M1
D1	5mm red LED	S1
R1	220Ω, 0.125W resistor	R1
	Jumper wires	H1
	Solderless breadboard	H2

LEDs are *polarized*, which means they will only work in the way you hook them up. The positive lead is called *anode*, which is the longest leg on an LED, and the negative lead is called *cathode*, which is recognized by a shorter leg. Also, if you look at the casing of the LED there should be a flat side on the lip, which also indicates the cathode side of the LED. LEDs are

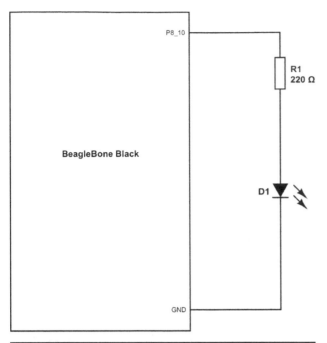

Figure 2-1   Schematic diagram for Project 2

also known as diodes or light-emitting diodes, and they only allow current to flow in one direction—from positive to negative. See Figure 2-1 for a schematic diagram of the LED circuit.

The resistor value is given based on the technical characteristics of the particular LED and using Ohm's law, which is probably the most important equation you will need to know. Ohm's law is the relationship between voltage (V), current (I), and resistance (R) in an electrical circuit. If we have the value of two of these fields, we can easily identify the third value using Ohm's law. LEDs have a predetermined voltage drop across them and are specifically designed to operate using a certain current value. The larger the current value of the LED, the brighter the LED will be up to a certain limit. You can find these values and technical information on the LED's datasheet, which can be supplied by the manufacturer or the store where you purchased the LED. See Chapter 8 for more information about reading datasheets.

With this information we can now calculate the value of the resistor we will need for this

project. We know that the supply voltage from the BeagleBone Black's GPIO pins is 3.3V, the forward voltage of the LED is 2V, and the current rating is 6mA (this is the maximum current value the BeagleBone Black's GPIO pins can provide). So using Ohm's law for solving what the resistor value is, we use the equation

$$R = 3.3V - 2V/0.006A$$

which gives us a maximum resistance value of 216Ω.

CAUTION   Using a lower-value resistor may cause damage to the BeagleBone Black by drawing too much current.

As you increase the value of the resistor, less current flows through the LED. Therefore, the LED is less bright but will nonetheless light up. For this project we use a 220Ω resistor, which is the closest fixed value of resistor that can be purchased.

Another very useful electrical equation is to calculate the power rating of the resistor that's required. For this particular project the power rating will be very low, so we can use a 0.125W- or 0.25W-rated resistor, but for larger projects that require brighter, more powerful LEDs, you will need more powerful resistors; otherwise, the resistors will generate too much heat and burn out. The power rating calculation is $P = I \times V$, where P is equal to power, I is the current rating, and V is voltage. Using the values we have for our LED circuit, we get

$$\Pi = 0.006A \times 3.3V - 2V$$

which is equal to 7.8mW. Therefore, we can safely say that these resistors will be fine for our circuit.

Connect your LED using the diagram in Figure 2-2. You can see that the anode leg of the LED is connected to the resistor and that

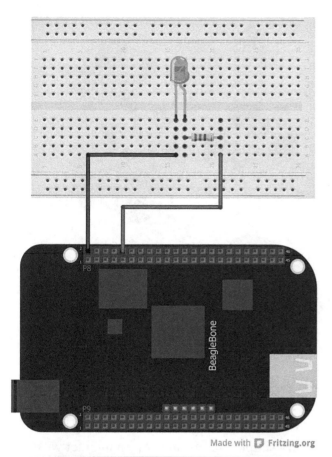

the cathode is connected to ground on the BeagleBone Black. Insert one of the jumper wires into the P8 header of the BeagleBone Black using either pin 1 or 2, as this is our ground connection. Connect the other end in the same rail as the cathode of the LED. To avoid confusion, it is always best to use black wires for ground connections and red wires for plus voltage. You can see the complete project in Figure 2-3.

Software

For this project we will be using the same code as in Project 1, but we will be making some small changes so we can use the GPIO pins on the BeagleBone Black board.

The JavaScript code for blinking an external LED follows next.

Figure 2-2   Breadboard layout for Project 2

Figure 2-3   Project 2 blinking external LED

```
var b = require('bonescript');
var led = "P8_10";

b.pinMode(led, b.OUTPUT);

var state = b.LOW;

b.digitalWrite(led, state);

setInterval(toggle, 1000);

function toggle() {
    if(state == b.LOW) state = b.HIGH;
    else state = b.LOW;
    b.digitalWrite(led, state);
}
```

Change the variable `led` value to `P8_10` (this is the GPIO pin that we will be using to control the external LED). The rest of the code will remain the same. Run the code in the Cloud9 IDE, and you will see the LED will blink on and off every second.

Not amazing, I know, but nonetheless this is our "Hello World" project—your introduction to the world of BeagleBone Black. We all have to start somewhere.

Project 3
Adjustable LED Blinker

Project 3 uses the analog pins on the BeagleBone Black to control the brightness of an LED using pulse width modulation (PWM). We will use PWM more in the chapters on using motors and sensors. We add two switches in this project, which are used to make the LED brighter or dimmer. This project utilizes analog and digital pins together and also shows you how to create functions and loops in JavaScript. Figure 2-4 shows a range of different size tactile switch which all work in the same way using switch gates.

Figure 2-4 Range of different tactile switches

You can see in Figure 2-5 that wiring a switch is very simple; this allows you to interact with the project in real time.

Pulse Width Modulation

Before we can start, you need to learn a little bit about PWM and how it works. PWM stands for pulse width modulation, and it is used to replicate an analog signal using a digital type of signal. The BeagleBone's PWM pins are controlled using an OMAP timing processor chip, which means that using the PWM pins in the BeagleBone Black utilizes none of the main

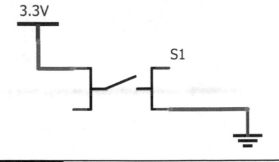

Figure 2-5 Schematic diagram for Project 3

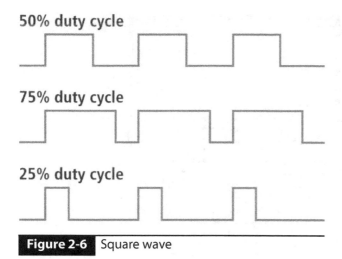

50% duty cycle

75% duty cycle

25% duty cycle

Figure 2-6 Square wave

CPU. PWM allows us to control electronics such as LEDs and motors by generating a high or low signal. PWM works by modulating the duty cycle of a square wave, which is a signal that switches on and off (see Figure 2-6).

The duty cycle is a percentage of time that a square wave is high or low. The analogWrite() function sets the duty cycle of a square wave depending on the value you set. Using a value of 0 will indicate a duty cycle of 0 percent, which is always low. In contrast, using a value of 1 will set the duty cycle of 100 percent, which is always high. As you might have guessed, a value of 0.5 will set the duty cycle to 50 percent. The best way to describe using the LED in our project is to think of a flip book with 50 pages: the pages in the book are either red (high) or white (low). If we set the duty cycle to 50 percent, then every other page of the book will be red, and when we flip through the book we see the red pages but the color looks dim when the pages are flipped quickly. Now if we set the duty cycle to 75 percent, then roughly three pages out of every four pages are red and one is white. When we flip through the book, we see the color red more predominantly. Therefore, as we increase or decrease the duty cycle, the LED gets brighter or dimmer. The flipping of pages is the frequency, and the pages themselves represent the duty cycle.

TABLE 2-2	Components and Equipment for Project 3	
Schematic Reference	**Description**	**Appendix A**
	BeagleBone Black	M1
D1	High-powered LED	S2
R1	110Ω resistor	R2
R2, 3	Two 4.6K resistors	R3
S1, 2	Two tactile switches	H3
	Solderless breadboard	H2
	Jumper wires	H1

Hardware

Table 2-2 lists the components and equipment required for Project 3. The hardware for this project is still very simple, but with the addition of a few switches and some pull-down resistors.

The switches are easy to use and wire up to a circuit using any standard tactile push switch with either two or four pins, which all serve the same function: to make or break a circuit connection. Basically, we are making the connection and then detecting whether a circuit has been made or a voltage is flowing through the circuit and then returning a digital value of either a 1 or 0. See Figure 2-7 for the breadboard layout diagram, and you can see the complete project in Figure 2-8.

Software

The JavaScript code for Project 3 is as follows:

```
var b = require('bonescript');
var awValue = 0.01;
var awDirection = 1;
var awPin = "P8_13";

b.pinMode(awPin, b.OUTPUT);
b.pinMode('P8_19', b.INPUT);
b.pinMode('P9_14', b.INPUT);
setInterval(check,100);

function check(){
b.digitalRead('P8_19', checkButton);
```

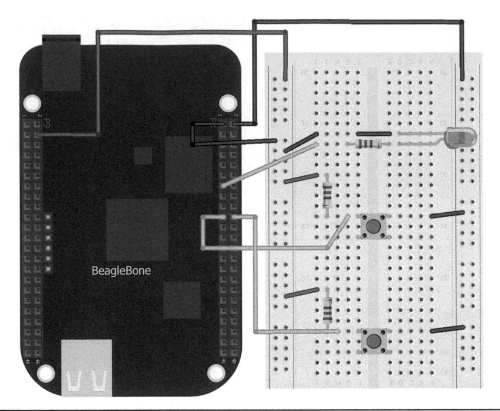

Figure 2-7   Breadboard layout for Project 3

Figure 2-8   Project 3 adjustable LED fader

```
b.digitalRead('P9_14', checkButton2);
}
function checkButton(x) {
    console.log (x.value);
    console.log (awValue);
        if(x.value == 0){
```

```
            b.analogWrite(awPin,
                awValue);
            awValue = awValue +
                (awDirection*0.01);
            if (awValue > 1) {
                awValue = 1 + (-1*0.01);
            }
        }
    }
function checkButton2(x) {
    console.log ('test' + x.value);
    console.log (awValue);
        if(x.value == 0){
            awValue = awValue +
                (-1*0.01);
            b.analogWrite(awPin,
                awValue);
            if (awValue <= 0.01) {
                awValue = 0 + (1*0.01);
            }
//          awValue = awValue +
                (awDirection*0.01);
        }
    }
```

The first part of the code is used for setting up the variable values. The `awvalue` variable indicates the value we will use to send to the LED to either make it brighter or dimmer, which will always be the value we will send to `analogWrite`. The variable `awDirection` is used to calculate the direction of `awValue` (that is, whether we will be increasing or decreasing the brightness of the LED).

```
var b = require('bonescript');
var awValue = 0.01;
var awDirection = 1;
var awPin = "P8_13";
```

We need to make sure we define whether the pins on the BeagleBone Black are inputs or outputs:

```
b.pinMode(awPin, b.OUTPUT);
b.pinMode('P8_19', b.INPUT);
b.pinMode('P9_14', b.INPUT);
```

If we think about the circuit we have and the components we are using, we can determine that the high-powered LED is an output because it emits light out and the two switches are inputs because we are detecting an input voltage from these switches when a connection has been made once they are pressed. You will notice the first line `bpinMode(awPin, b.OUTPUT);` does not refer to a GPIO pin but to a variable. This is an alternative way to reference the GPIO pin, and it allows us to name the pin anything we want so that it's easier to call back to it in our source code.

For our program to run, we first need to listen for the state of the switch, which will either be pressed down (on) or in a normal state (off). If the state of the switch is off, nothing is going to happen. However, we need to keep checking the state of the switch for when we actually do press it (so that it's "on") so that we can run a function. To do this, we can create a loop that will keep running in intervals of the time we set:

```
setInterval(check,100);
```

This code sets an interval of every 100 milliseconds and will run a function called `check`. This allows us to check for a button press in the function within an amount of time that's likely to catch it. If the time interval was set to 500 milliseconds, for example, there is a possibility we could miss the button press.

NOTE It's always good to play around with the timings to find a suitable interval that works for your project.

Now that we have set the interval to call a function, we need to write the function code that will check the status of our buttons:

```
function check(){
b.digitalRead('P8_19', checkButton);
b.digitalRead('P9_14', checkButton2); }
```

In this function we are reading the state of the buttons `P8_19` and `P9_14` and outputting the value to the functions `checkButton` and `checkbutton2`, respectively. Now that we have the button state, we can check whether or not the button pressed down by checking whether the value is equal to 1 or 0 using an `if` statement:

```
function checkButton(x) {
    console.log (x.value);
    console.log (awValue);
        if(x.value == 0){
            b.analogWrite(awPin,
              awValue);
            awValue = awValue +
              (awDirection*0.01);
            if (awValue > 1) {
                awValue = 1 + (-1*0.01);
            }
        }
    }
```

If the state of the button is high, we want to write a value to the LED to light it up. We use `analogWrite` to light up the LED using the `awValue` we set at the beginning. After we send the value, we want to increase it incrementally so that next time the button is pressed, we send a higher value, thus gradually increasing the brightness of the LED. To do this, we add the calculation

```
Awvalue = currentawValue +
   (awDirection*0.01)
```

where `awDirection` determines whether we add or subtract the `awValue`, which will, in turn, make the LED brighter or dimmer, respectively.

In the function we also need to set a limit to the value that we write to the LED; otherwise, we could see an infinite value that cannot be written, and we will either damage the LED or see an error in our program. Our value limit is equal to 1. Therefore, we write an if statement to say that if the `awValue` is greater than 1, we reset that value back to the previous value written so when the push button is pressed, it will just send the same (highest) value to the LED and thus not exceed that value. This uses an equation where `awValue` is equal to the following:

$$1 + (-1*0.01)$$

This equation subtracts 0.01 from the current `awValue` and gives us a safe value to write to the LED.

In the function `checkButton2`, we check the state of the second button and write code that's similar to `checkButton` to dim the LED:

```
function checkButton2(x) {
    console.log ('test' + x.value);
    console.log (awValue);
        if(x.value == 0){
            awValue = awValue +
              (-1*0.01);
            b.analogWrite(awPin,
              awValue);
            if (awValue <= 0.01) {
                awValue = 0 + (1*0.01);
            }
        }
    }
}
```

The code is the same except that we change the direction of the value by writing a negative value using `awDirection`. This will decrease the overall value and dim the LED as we push the second button.

Project 4
High-Powered LED Morse Code Sender

Morse code is a form of communication for sending messages using dots and dashes. Morse code has been around since the nineteenth century and can be used in a variety of ways, such as radio waves and light signals. Morse code is commonly used as an SOS (save our souls) distress signal. This signal is known all around the world as the sequence

...---...

and was introduced over 100 years ago.

In this project we will create a high-powered LED circuit that uses a transistor to make our LED flash the sequence of SOS over and over again, thus creating an Evil Genius distress beacon.

> **NOTE** Do not look directly into the LED for long periods of time, as this may cause damage to your eyes.

This chapter requires a bit of soldering. We'll make a prototype shield for the BeagleBone Black that we can plug in directly to the top of the board; this is ideal for embedding into enclosures or cases and keeps our projects discreet and compact if space is at a minimum.

Hardware

Table 2-3 lists the components and equipment required for Project 4. The hardware for this project is similar to those components used in

TABLE 2-3	Components and Equipment for Project 4	
Schematic Reference	**Description**	**Appendix A**
	BeagleBone Black	M1
D1	High-powered LED	S2
R1	30R resistor	R4
R2	4.7R resistor	R5
T1	BD139 power transistor	S3
	Solderless breadboard	H2
	Jumper wires	H1
	Protoshield kit (optional)	M4

Project 2, with the introduction of using an NPN transistor.

We need to use this transistor because the BeagleBone Black can only output a maximum of 6mAh of current; however, our high-powered LED requires up to 70mAh, which is a lot of current. At the end of Chapter 1 we talked about calculating the amount of power required using the following formula:

$$P = IV$$

The power (P) is equal to the voltage (V) across something multiplied by the current (I) flowing through it, and the unit of power output is equal to the wattage. So with this in mind, we can work out that the maximum power output from one of the BeagleBone Black GPIO pins is equal to approximately 20mW; for Project 2 this was more than enough power to drive our LED, but for this project we are driving a high-powered LED up to 70mAh, so our BeagleBone Black cannot directly drive this LED. We need to somehow amplify the current.

This is a common formality within electronics, and you will encounter this issue many times with different projects requiring high power to run them, such as when you work with motors. The theory is simple in that we start off with a small current from the BeagleBone Black GPIO pins and we need to somehow amplify that

current. A small component that is commonly used for achieving this effect is the transistor. We will use a transistor in our circuit to switch on and off our high-powered LED with the amplifying of the current.

A transistor is a simple device, and a very useful one at that. Currently, there are many different types of transistors that come in all different shapes and sizes, each one having a varying function. In Figure 2-9, you can see that they come as small as the eye can see. This type is also known as a surface mount device (SMD) transistor, and it requires special soldering using high-tech professional equipment. There are also some larger transistors with flanges and some that require a heat sink for heat dissipation.

In this project we are going to use a transistor called an NPN bipolar transistor; this is the most commonly used type in electronic circuits. This transistor has three legs: the collector, the emitter, and the base (see Figure 2-10). Basically, a small current flows through the base and is amplified to a much larger current and will flow between the collector and the emitter. See Figure 2-11 for a schematic diagram of Project 4.

The amplification of our circuit depends on a couple of factors, one being the specification of the NPN transistor. After checking the technical datasheet of the BD139 transistor, we can see

Figure 2-9 A range of different-size transistors

Figure 2-10 BD139 transistor

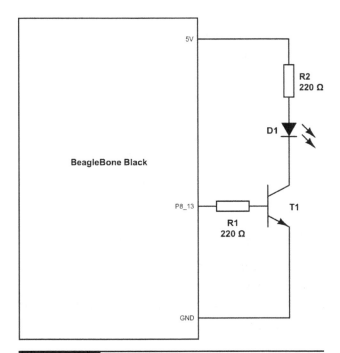

Figure 2-11 Schematic diagram for Project 4

that we can have a maximum input current of 0.5A and we can amplify this current up to a maximum of 1.5A. Therefore, if we use the 220R resistor that we used in Project 2 to drive the LED up to 6mA, we could typically expect our NPN transistor to amplify this up to 60mAh, working on an amplification factor of 100, which is typically common in NPN transistors. This will be more than enough to switch more power through our high-powered LED.

The datasheet for our NPN transistor says that we can have a maximum forward current of 250mA and a forward voltage of 2.6V. So with this in mind, we will aim to amplify the current up to 60mA, which will make the LED nice and bright and won't increase the risk of shortening its operational life.

R2 limits the current flowing through the LED to around 60mA. Therefore, using Ohm's law, we can calculate the value of our resistor using R = V/I. The voltage (V) will be 5 (supply voltage) – 2.6 (LED forward voltage) – 0.6 (between the base and the emitter transistors a voltage drop of 0.6V normally occurs, so we must subtract this value from our equation). So R = 1.8/0.06 gives us a resistive value of 30R. We must also make sure we select a resistor that can cope with the required power rating as power is dissipated through the resistor as heat. Therefore, 60mA × 1.6V = 96mW. This means we can safely use a 0.125W resistor or higher. Figure 2-12 shows the breadboard layout for Project 4. The completed Project 4 can be shown in Figure 2-13.

When creating projects, usually a breadboard and some jumper wires will suffice, but when you have tested your circuit design, you may want to build a more stable solution for embedding your project. Figure 2-14 shows Project 4 soldered onto a prototyping cape PCB, this allows the project to be more permanent.

Software

The JavaScript code for Project 4 is as follows:

```
var b = require('bonescript');

var dot = 200;
var dash = dot * 3;
var pause = dot;


b.pinMode('P8_13', b.OUTPUT);
```

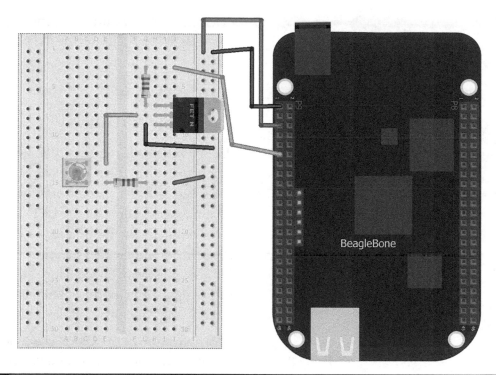

Figure 2-12   Breadboard layout for Project 4

Figure 2-13   Project 4 high-powered Morse code sender

Figure 2-14 Morse code sender expansion cape

```
function ledsOn() {
    b.digitalWrite('P8_13', 1);
}

function ledsOff() {
    b.digitalWrite('P8_13', 0);
}

function flash(period) {
    ledsOn();
    setTimeout(ledsOff, period);
}

function sequence(durations) {
    var timeline = 0;
    for (var i = 0; i < durations.
length; i++) {
        var d = durations[i];
        setTimeout(function() { flash(d);
}, timeline);
        timeline = timeline + d + pause;
    }
}

sequence([ dot, dot, dot, dash, dash,
dash, dot, dot, dot]);
```

This program uses a series of functions to write the state of the LED to 1 or 0. The functions are then used in the sequence to play a series of flashes. The flashes of the LED are determined by the duration period set in the variables at the beginning of the program.

Project 5
RGB LED Fader

Project 5 uses three analog 10K potentiometer sliders to control the color of an RGB LED (see Figure 2-15). Each slider is used to control the brightness of a color on the LED: either red, green or blue. When the colors are combined in different variations, they can create other colors. This is a great way of reducing the amount of LEDs required in a project and also gives your project more visibility and flexibility.

A potentiometer (POT) is essentially a three-terminal resistor that acts as a voltage divider. When you adjust the value of the resistor, the voltage is increased or decreased in turn. Potentiometers are more commonly used in electronics for volume control and come in either a linear taper or logarithmic design. Our slider potentiometer is a logarithmic POT, and this means that the resistance tapers from one end to the other, so the output voltage is dependent on the slider position (see Figure 2-16).

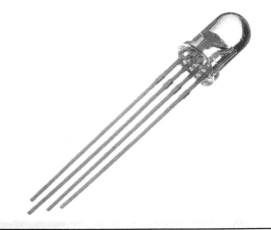

Figure 2-15 RGB LED (10mm)

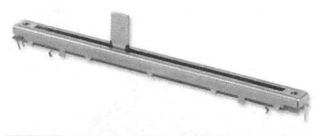

Figure 2-16 Slider potentiometer

Hardware

The hardware for this project is very simple: it consists of a 10mm standard RGB LED and some 10K slider potentiometers. The RGB LED has four legs, three of which are used to control the colors red, green, and blue. These are connected to the BeagleBone Black just as you would connect a standard LED (like in Project 2). The longest leg of the LED is the common cathode pin, and this is connected to the ground pin on the BeagleBone Black. The three 10K

TABLE 2-4	Components and Equipment for Project 5	
Schematic Reference	**Description**	**Appendix A**
	BeagleBone Black	M1
D1	RGB LED	S3
R1	220R resistors	R1
R2, 3, 4	10K potentiometer	R4
	Solderless breadboard	H2
	Jumper wires	H1

POTs are connected to the BeagleBone Black analog input pins and can read the voltage value from 0V to 1.8V (max); the value is converted to 0 (0V) or 1 (1.8V). Table 2-4 lists the components and equipment required for Project 5.

The breadboard layout for Project 5 can be seen in Figure 2-17 and the final completed project in Figure 2-18.

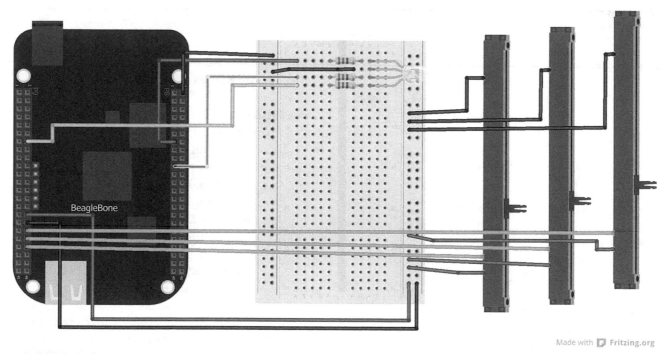

Made with [□] Fritzing.org

Figure 2-17 Breadboard layout for Project 5

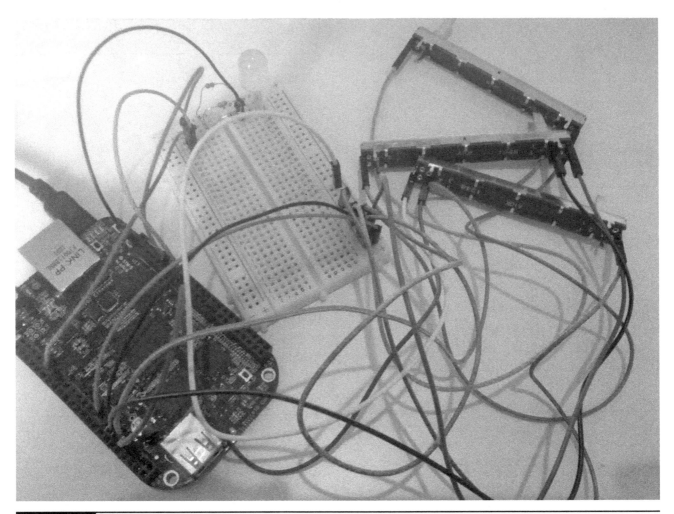

Figure 2-18 Project 5 RGB LED fader

Software

The sketch for this project uses the `analog.Read` function to read the current value of the 10K POT sliders.

Here is the JavaScript code we will use for Project 5:

```
var b = require('bonescript');

var RedPin = "P8_13";
var GreenPin = "P8_19";
var BluePin = "P9_14";

var S1Value = 0.0;
var S2Value = 0.0;
var S3Value = 0.0;

b.pinMode(RedPin, b.OUTPUT);
```

```
b.pinMode(GreenPin, b.OUTPUT);
b.pinMode(BluePin, b.OUTPUT);

setInterval(fade, 100);
function fade(){
    b.analogRead('P9_40', Red);
    function Red(x) {
        S1Value = x.value;
        b.analogWrite(RedPin, S1Value);
        console.log('Red value = ' +
          x.value);
        console.log('x.err = ' + x.err);
        b.analogRead('P9_38', Green);
            function Green(x) {
                S2Value = x.value;
                b.analogWrite(GreenPin,
                  S2Value);
                console.log('Green value
                  = ' + x.value);
```

```
                     b.analogRead('P9_36',
                Blue);
                        function Blue(x) {
                            S3Value =
                              x.value;

   b.analogWrite(BluePin, S3Value);
                            console.log
                            ('Blue value =
                             ' + x.value);

     }}}}
```

The first part of the code is where we set our variables. We give each pin of the LED a variable so we can use `analogWrite` to write the value of the LED color; we name these variables the color of the LEDs we are changing to make things easier. We also set a variable for the value that is read from the 10K POTs: slider 1 (S1), slider 2 (S2), and slider 3 (S3). We set these values to zero to start off with so all the LEDs are in an off state.

```
var RedPin = "P8_13";
var GreenPin = "P8_19";
var BluePin = "P9_14";
```

Once the variables have been set, we need to state whether the pins are an input or an output. The analog pins we are using for the slider are automatically set for analog input, so we only need to set the LED pins to OUTPUT:

```
b.pinMode(RedPin, b.OUTPUT);
b.pinMode(GreenPin, b.OUTPUT);
b.pinMode(BluePin, b.OUTPUT);
```

To make sure we are always reading the status of the 10K POTs, we need to create a loop so that we can read the values for the amount of time for which we set the loop. We use the `SetInterval` function to create a loop that will run every 100 milliseconds; this will make the LED change almost in real time or as soon as we slide the potentiometer. For the purpose of running the program for the first time, you may want to change the interval to 1,000 milliseconds just to slow things down a little bit. This way,

you can see the output values being written to the LED.

```
setInterval(fade, 100);
```

Once the loop function is called, we then read the status of the first slider, which controls the red LED on our RGB LED. Therefore, we use the `analogRead` function to read the status of the slider and output the value to another function, which we will use to write the value to the LED.

```
b.analogRead('P9_40', Red);
   function Red(x) {
       S1Value = x.value;
       b.analogWrite(RedPin, S1Value);
       console.log('Red value = ' +
         x.value);
```

You can see in this code that we read the value from analog pin `P9_40` and we output this to a function called `Red`. In the `Red` function, we set `S1Value` to the value of the slider. Because the PWM pin writes a value from 0 to 1, which in turn makes the LED brighter or dimmer, we can set the value the same way; therefore, we do not need to create an equation to convert the value. Once we have our value, we write this to the red LED pin using `analogWrite`. You will also notice that we use the console log to output the value in the console; it is always a best practice to use this for debugging your project and making sure the correct values are being written to the LED. Once we have written the value for the red LED, we then move on to the next color LED, which is green. Therefore, all we do is repeat the process of the function and change all the variables to the green LED:

```
b.analogRead('P9_38', Green);
           function Green(x) {
           S2Value = x.value;
           b.analogWrite(GreenPin,
             S2Value);
           console.log('Green value
             = ' + x.value);
```

We repeat the process for the final (blue) pin on the LED, reading the slider value and writing the value to the LED.

> **NOTE** Make sure when you create a function that you close that function using the right brace (}).

Project 6
Traffic Lights

In Project 6, we are going to build a model of a traffic light system using red, amber, and green LEDs. Every time we press the button, the traffic light will change to the next sequence. In the United Kingdom and many other European countries, the traffic light signal goes from red to red/amber, green, green/amber, and then back to red.

Hardware

The components for this project are listed in Table 2-5. If possible, it is always a best practice to choose the same specification of LEDs (usually the same manufacturer brand or series). This will give us the same brightness for all the lights and also allow us to use the same value of resistors for the LEDs.

TABLE 2-5	Components and Equipment for Project 6	
Schematic Reference	Description	Appendix A
	BeagleBone Black	M1
D1	5mm red LED	S1
D2	5mm amber/yellow LED	S4
D3	5mm green LED	S5
R1, 2, 3	Three 220R resistors	R1
R4	10K pull-down resistor	R5
	Breadboard	H2
	Jumper wires M-to-F	H1
S1	Tactile switch	H3

The circuit diagram is shown in Figure 2-19 for Project 6. The LEDs are connected to the digital pins on the BeagleBone Black. As in Project 2, each LED has a current-limited resistor based on Ohm's law. Also, a tactile switch is connected to a digital input pin. When the switch is connected or pushed down, it makes a circuit and a voltage passes through that is detected by the digital input. The completed project is shown in Figure 2-20.

Software

Here is the JavaScript code for Project 6:

```
var b = require('bonescript');

var red = "P8_13";
var yellow = "P8_15";
var green = "P8_17";
var inputPin = "P8_10";

var switchstatus =0;

b.pinMode(red, b.OUTPUT);
b.pinMode(yellow, b.OUTPUT);
b.pinMode(green, b.OUTPUT);
b.pinMode(inputPin, b.INPUT);

var on = b.HIGH;
var off = b.LOW;

setInterval(check,200);

function check(){
b.digitalRead(inputPin, checkButton);
}

function checkButton(x) {
    if (x.value == 1){
        switchstatus++;
        switch (switchstatus){
            case 1: setTimeout (function
                (){b.digitalWrite(red,
                on);
                b.digitalWrite(yellow, off);
                b.digitalWrite(green,
                off);}, 500);
                break;
```

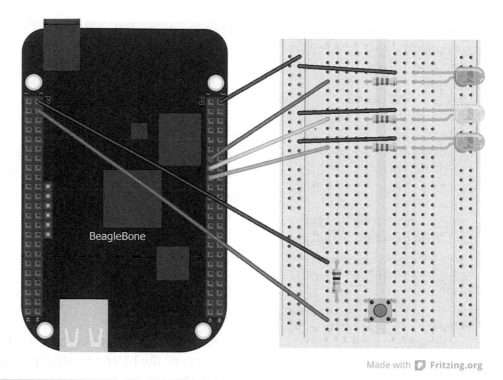

Figure 2-19　Breadboard layout for Project 6

Figure 2-20　Project 6 traffic light system

```
case 2: setTimeout (function
  (){b.digitalWrite(yellow,
  on);}, 500);
    break;
case 3: setTimeout (function
  (){b.digitalWrite(red,
  off);
b.digitalWrite(yellow, off);
  b.digitalWrite(green,
  on);}, 500);
    break;
case 4: setTimeout (function
  (){b.digitalWrite(yellow,
  on);
b.digitalWrite(green, on);},
  500);
    break;
default: switchstatus = 0;
    break;
    }
  }
else if (switchstatus >=5){
    switchstatus = 0;
  }
}
```

The code for this project is very simple: we only check to see if the switch is pressed and then we run the script in the traffic light sequence. You can see that we created a loop so that we can always check to see if the switch has been pressed. Each sequence runs for a delayed amount of time; otherwise, the LEDs would blink so fast that we would never see them. This is why we set a timeout. The first part of the code sets the variables for each LED. We use a digital output pin for each LED and a digital input pin for the switch.

```
var red = "P8_13";
var yellow = "P8_15";
var green = "P8_17";
var inputPin = "P8_10";

var switchstatus =0;

b.pinMode(red, b.OUTPUT);
b.pinMode(yellow, b.OUTPUT);
```

```
b.pinMode(green, b.OUTPUT);
b.pinMode(inputPin, b.INPUT);
```

We also set a variable to check which part of the LED traffic light sequence comes next, and every time we change the sequence, the switchstatus increases by 1. In order for us to check to see if the switch has been pushed down, we need to read the status of the switch all the time. Therefore, we set a loop to check every 200ms the status of the switch and output the value to a function called checkButton:

```
setInterval(check,200);
function check(){
b.digitalRead(inputPin, checkButton);
}
```

We now create the checkButton function. If the value of the button is equal to 1, we know that the button has been pressed and we can initiate our switch statement. Before we enter the switch statement, we increase the switch status by 1 so that every time we run the program, it will increase by 1—and more importantly, switch to the next statement in the sequence.

```
function checkButton(x) {
    if (x.value == 1){
        switchstatus++;
```

The switch statement is very simple: it reads the value after switch, which in this case is called switchstatus, and whichever value it is, it switches to that case and runs the code in that case. The switch cases are numbered 1 to 4, and also feature a default value so that if the switchstatus value does not equal any of the case numbers, the default case will be used. Once we have switched to a case or sequence, we write our LED sequence; we do this by setting a timeout function to add a short delay of 500ms, using the digitalWrite function, and calling the color of the LED we want and stating whether it should be on or off.

```
switch (switchstatus){
   case 1: setTimeout (function ()
            {b.digitalWrite(red, on);
   b.digitalWrite(yellow, off);
     b.digitalWrite(green, off);}, 500);
       break;
   case 2: setTimeout (function ()
     {b.digitalWrite(yellow, on);},
      500);
       break;
   case 3: setTimeout (function ()
            {b.digitalWrite(red, off);
   b.digitalWrite(yellow, off);
       b.digitalWrite(green, on);},
       500);
       break;
   case 4: setTimeout (function ()
            {b.digitalWrite(yellow, on);
   b.digitalWrite(green, on);}, 500);
```

The default value in the `switch` statement will always be used when the `switchstatus` value is not equal to 1, 2, 3, or 4, in which case we reset the value back to 0 so we can start the sequence again.

```
default: switchstatus = 0;
                break;
```

NOTE Make sure you add **break;** at the end of each case; otherwise, the code will run the next line.

Run the sketch program and push the button to start the first sequence of lights. Continue to press the button until you get back to the first sequence of lights again.

Project 7
Matrix Displays

LED arrays and dot matrix displays are great components for getting across messages in either text or image format—they're also great for creating some Evil Genius eyes. LED matrix displays usually consist of an array of LEDs in rows and columns. Typically, these come in 4×4, 8×8, and 16×16 for small hobbyist projects, or you can even create your own matrix using LEDs. Most matrix displays come in a single color of LED, but bicolor LEDs are also available. These are great for creating a range of different colors to give a great visual effect for some projects.

This project uses a matrix LED module that has a backpack on it that includes a HOLTEC HT16K33 I2C IC. This reduces the amount of wires needed to connect to the BeagleBone Black. A standard matrix display with an 8×8 array has up to 24 pins, and if we connected all these pins to the BeagleBone Black, this would take up almost half of the GPIO pins and would not leave us many more pins for adding other devices.

Using I2C

I2C was designed by Philips in the early 1980s. It was designed to provide an easy form of communication between electronic devices that are on the same circuit. I2C has some great features; for example, only two wires or connections are required, and it operates using a master/slave network. Each I2C device that is connected has its very own unique address that is used to identify the components on a circuit. In terms of our project, this protocol is ideal for the matrix display; it means that we only need two single wires to connect to our BeagleBone Black, and we can add as many as 127 different devices.

Some devices have an address that is fixed by the manufacturer; others can be configured to take an address from a range of addresses. When a device is used as a slave, it is normally possible to configure its address via software. The address may be followed by one or more bytes of data that go from master to slave or from slave to master.

When data is being sent on the SDA line, clock pulses are sent on this line to keep the master and slave synchronized. When the devices are communicating, they only send one bit at a time, so this means the data transfer rate is one-eighth of the clock rate. The BeagleBone Black has two I2C busses and can be set at variable speeds.

Hardware

Table 2-6 lists the components and equipment required for Project 7.

TABLE 2-6	Components and Equipment for Project 7	
Description	**Appendix A**
BeagleBone Black	M1
LED matrix display	S4
Solderless breadboard	H2
Jumper wires	H1

The LED matrix in this project comes as a kit (see Figure 2-21). This will require some beginner-level soldering because the matrix display needs to be soldered onto the backpack (PCB) board, and a single row of four-way headers also needs soldering to attach to our breadboard. When you solder the LED display to the PCB board, it is very important that you make sure it faces the right way; otherwise, the pinout will be incorrect and may damage the board and the LED itself. If this is the case, you

can easily desolder the LED and reposition it—but it's always best to do it right the first time.

You can see in Figure 2-22 that we connect the power pins (+5V and GND) to the matrix display. The other two pins are labeled "SDA" and "SCL," and these are connected to the BeagleBone's I2C pins SDA and SCL. The SDA pin is used to send and receive the data sent between devices, and the SCL pin is used to synchronize the communications. The completed project can be seen in Figure 2-23.

Software

Before we start, we need to make sure the I2C matrix display is connected correctly to the BeagleBone Black and that communication between the devices can be established. To do this, we can use the built-in I2C tools that come

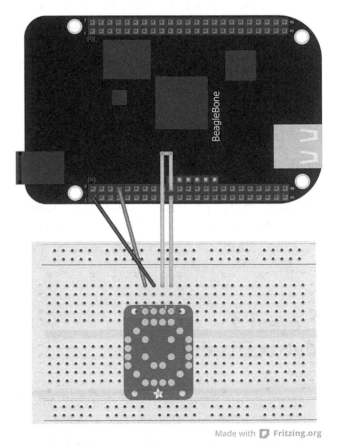

Figure 2-22 Breadboard layout for Project 7

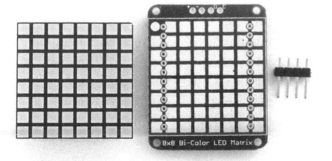

Figure 2-21 The Adafruit LED matrix module kit

Figure 2-23 Project 7 matrix display

with the BeagleBone Black board; these tools are used to provide useful information such as a device's I2C address. To issue the commands, you need to open up a Terminal window either through GateOne SSH or a program such as PuTTY or ZOC Terminal (see Chapter 1 for more information on connecting to your BeagleBone Black through SSH). Once you're in the Terminal window, type the following command:

```
I2cdetect —y —r 1
```

This command is used to detect devices on the I2C bus and display their device address. You can see in Figure 2-24 that we have detected an address of 0x70, which is our LED matrix board. You may also see the value UU, which means the address is busy (usually on the BeagleBone Black such addresses are taken up by the cape EEPROMs).

```
     0  1  2  3  4  5  6  7  8  9  a  b  c  d  e  f
00:          -- -- -- -- -- -- -- -- -- -- -- -- --
10: -- -- -- -- -- -- -- -- -- -- -- -- -- -- -- --
20: -- -- -- -- -- -- -- -- -- -- -- -- -- -- -- --
30: -- -- -- -- -- -- -- -- -- -- -- -- -- -- -- --
40: -- -- -- -- -- -- -- -- -- -- -- -- -- -- -- --
50: -- -- -- -- UU UU UU UU -- -- -- -- -- -- -- --
60: -- -- -- -- -- -- -- -- -- -- -- -- -- -- -- --
70: 70 -- -- -- -- -- -- --
root@beaglebone:~#
```

Figure 2-24 BeagleBone Black I2C addresses

Here is the JavaScript code for Project 7:

```
var i2c = require('i2c');
var address = 0x70;
var wire = new i2c(address, {device:
    '/dev/i2c-1'});
var brightnessArray = [0xE0, 0xE1, 0xE2,
    0xE3, 0xE4, 0xE5, 0xE6, 0xE7];
var smiley = [0x1e,0x21,0xd2,0xc0,0xd2,
    0xde,0x21,0x1e];
var brightness = brightnessArray[1];

    wire.writeBytes(0x21, 0x00);
    wire.writeBytes(0x81, 0x00);
    wire.writeBytes(brightness, 0x00);

    wire.writeBytes(0x00,
        [smiley[1]);     // Row 1
    wire.writeBytes(0x02,
        [smiley[2]);     // Row 2
    wire.writeBytes(0x04,
        [smiley[3]);     // Row 3
    wire.writeBytes(0x06,
        [smiley[4]);     // Row 4
    wire.writeBytes(0x08,
        [smiley[5]);     // Row 5
    wire.writeBytes(0x0A,
        [smiley[6]);     // Row 6
    wire.writeBytes(0x0C,
        [smiley[7]);     // Row 7
    wire.writeBytes(0x0E,
        [smiley[8]);     // Row 8
```

The first line of our code is where we access all the I2C library functions, just like in our other projects where we reference the BoneScript library:

```
var i2c = require('i2c');
```

In order for us to send commands to the matrix display, we need to first point to the I2C address using a variable. The first line is the I2C address; this is the address we found earlier by using the I2C tools command in the Terminal window (0x70). The next command is the I2C address directory on the BeagleBone Black:

```
var address = 0x70;
var wire = new i2c(address,
    {device: '/dev/i2c-1'});
```

In order to set the brightness how we want it, we can create an array of brightness values ranging from 0 to 7 (0 being the dimmest and 7 being the brightest). An array is used to store data or objects in a variable and can easily be indexed or identified when called upon.

```
var brightnessArray = [0xE0, 0xE1, 0xE2,
    0xE3, 0xE4, 0xE5, 0xE6, 0xE7];
```

To light up our matrix display, we need to send eight bytes of data, for each row of the matrix display. If we want to light up all the LEDs on the first row, our byte of data would be the following binary value:

```
11111111
```

A value of 1 represents that the LED is on, and a value of 0 represents off. Now that we have our first byte of data we need to convert this into hexadecimal so we can write it to the matrix display. Converting a binary value to hexadecimal is very simple. Our binary number 11111111 gets divided up into four bits (that is, 1111 and 1111). Each decimal digit in the binary number gets doubled as you move on to the next binary number, starting with 1. Therefore, 81412111 would be equal to 15 (decimal value) or F (hexadecimal value). Also, remember that we always read binary from right to left. Table 2-7 shows the binary conversion.

If we want to light up a smiley face on the matrix display, we would set each LED to either a 1 or a 0 (see Figure 2-25). When we convert each row into hexadecimal, we get the following values:

```
var smiley = [0x1e,0x21,0xd2,0xc0,0xd2,
    0xde,0x21,0x1e];
```

As described previously, we created a variable to store the value of the brightness of the matrix display, and now we can set our variable to call using the following code:

```
var brightness = brightnessArray[1];
```

If we wish to change the brightness, we can change the value inside the parentheses from 1

TABLE 2-7	Binary Conversion	
Binary	Decimal	Hexadecimal
0000	0	0
0001	1	1
0010	2	2
0011	3	3
0100	4	4
0101	5	5
0110	6	6
0111	7	7
1000	8	8
1001	9	9
1010	10	A
1011	11	B
1100	12	C
1101	13	D
1110	14	E
1111	15	F

to 7. Before we can write the data to the matrix display, we need to make sure it's set up in order to receive the bytes of data. The first command we send is to set up the matrix display by putting it in standby mode, ready to receive the data. The second command sets the display on and

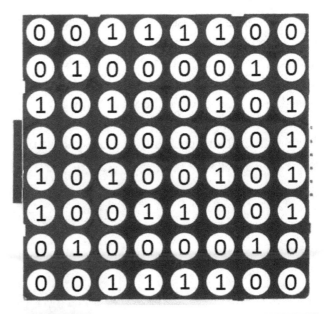

Figure 2-25   LED matrix binary code diagram

turns off any blinking so that the LEDs stay on all the time once we send the data. We send the brightness value as set in our variable earlier (we can easily play around with these values later on).

```
wire.writeBytes(0x21, 0x00);
            //system setup
wire.writeBytes(0x81, 0x00);
            //display on and no blinking
wire.writeBytes(brightness, 0x00);
            //set LED brightness
```

Now that we have everything set up, we can send the bytes of data to the LED matrix and create our smiley face. If we check the matrix display's LED chip driver datasheet, it tells us that the first byte to send is the row selection, which is 0x00. The next row is 0x02, and so on up to 0x0E, which is row 8. The next byte we send is the image value from our variable smiley, and because we want the first value in the variable, we use smiley[1]. We then cycle through the rest of the rows writing the next value in the variable.

```
wire.writeBytes(0x00,
            [smiley[1]);      // Row 1
wire.writeBytes(0x02,
            [smiley[2]);      // Row 2
wire.writeBytes(0x04,
            [smiley[3]);      // Row 3
```

```
wire.writeBytes(0x06,
            [smiley[4]);      // Row 4
wire.writeBytes(0x08,
            [smiley[5]);      // Row 5
wire.writeBytes(0x0A,
            [smiley[6]);      // Row 6
wire.writeBytes(0x0C,
            [smiley[7]);      // Row 7
wire.writeBytes(0x0E,
            [smiley[8]);      // Row 8
```

When you run the program, you should see a nice bright smiley face—not so evil, I know, but maybe you can create an Evil Genius face by playing around with some of the values.

Summary

This concludes our chapter on LEDs. Hopefully you have learned a lot about how you can use them in your projects. We used both analog and digital pins to show the varying effects they give us, and we also created different colors using RGB LEDs and images using matrix displays. Next, we are going to move on to create projects that sense input from the real-world environment around us.

CHAPTER 3

Sensor Projects

This chapter is all about sensing inputs from the real world around us. We can use this information to control outputs such as motors and lights, or we simply display the information. This chapter includes a variety of different sensor projects that could be used to create robotics, weather stations, flood warning systems, day/night switches, and location-based projects.

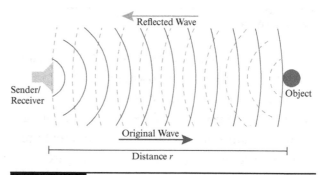

Figure 3-1 Ultrasonic sound waves

Project 8
Scanning Sonar

Scanning sonar using ultrasonic sensors is the most convenient way of sensing the distance of objects around us. An ultrasonic sensor generates high-frequency sound waves and receives back an echo of those sound waves. The sensor then calculates the time interval between sending the high-frequency sound wave and receiving the echo back, which is used to determine the object's distance, as shown in Figure 3-1.

Hardware

Table 3-1 lists the components and equipment required for Project 8.

The hardware for this project uses a MaxSonar EZ2 ultrasonic sensor, which is commonly found in hobbyists shops and

TABLE 3-1	Components and Equipment for Project 8	
Schematic Reference	**Description**	**Appendix A**
	BeagleBone Black	M1
	Ultrasonic sensor	M2
R1	330Ω resistor	R6
R2	220Ω resistor	R1
	Solderless breadboard	H2
	Jumper wires	H1

electronic stores. This sensor has three pins: V+, GND, and AN. The AN pin outputs an analog voltage with a scaling factor of Vcc/512 per inch. The sensor requires a 5V input from the BeagleBone Black. Because we are connecting the AN wire to the analog pins on the BeagleBone board to read the voltage input, we need to create a voltage divider using two resistors so that we do not damage the board.

V_{in}
Z_1
V_{out}
Z_2

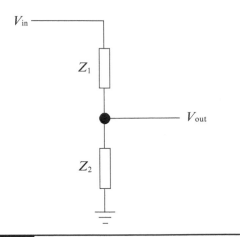

Figure 3-2 Voltage divider circuit

A voltage divider can be very useful in circuits when the input voltage is always higher than the output voltage required, such as using a 9V PP3 battery to power a 5V DC motor. A voltage divider usually consists of two resistors that are set up in a series circuit, as shown in Figure 3-2.

The voltage divider equation assumes that you know the three values of the circuit: the input voltage (V_{in}) and both resistor values (R1+R2).

$$V_{out} = V_{in} \bullet \frac{R_2}{R_1 + R_2}$$

Figure 3-3 Voltage divider equation

With these values, we can use the equation in Figure 3-3 to find the output voltage (V_{out}).

The equation clearly states that the output voltage is directly proportional to the input voltage and the addition of both resistors. Therefore, if our V_{in} is 5V and our R1 and R2 values are 330R and 220R, respectively, we can use this equation to calculate that the output voltage is 1.8V, which is what we want to read the voltage input from the analog pins.

The breadboard layout is shown in Figure 3-4. Make sure that both resistors are wired up correctly in the order shown in the diagram.

Software

Here is the code we'll use for Project 8:

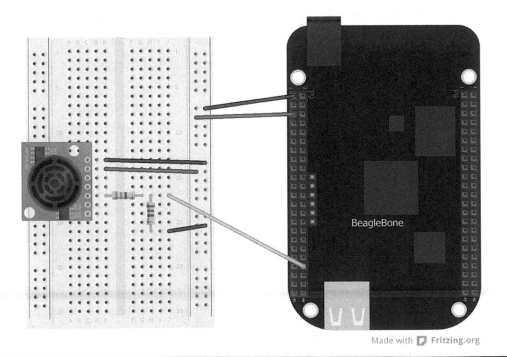

P9
P8
BeagleBone
Made with Fritzing.org

Figure 3-4 Breadboard layout for Project 8

```
var b = require('bonescript');
var analogVoltage = 0;

/* Check the sensor values every 2
   seconds*/
setInterval(read, 2000);

function read(){
    b.analogRead('P9_40', printStatus);
}

function printStatus(x) {
    var distanceInches;
    analogVoltage = x.value*1.8; // ADC
      Value converted to voltage
    console.log('x.value = ' +
      analogVoltage);
    distanceInches = analogVoltage /
      0.00699;
    console.log("There is an object " +
    parseFloat(distanceInches).
      toFixed(3) + " inches away.");
}
```

We first set up a loop so that we can check the status of the sensor every two seconds; we use the `setInterval` function to do this, and we run the function `read` every two seconds:

```
setInterval(read, 2000);
```

Once we have initiated the function, we then use the `analogRead` function to read the value of the sensor's output voltage and then output the value to a function called `printStatus`:

```
function read(){
    b.analogRead('P9_40', printStatus);
}
```

Once we have our analog voltage value, we can use this to calculate the distance using a simple mathematical equation in our code. The distance is equal to the analog voltage divided by 0.00699, which according to the datasheet, running at 5V input every inch is equal to 6.99mv of analog voltage:

```
function printStatus(x) {
    var distanceInches;
    analogVoltage = x.value*1.8; // ADC
      Value converted to voltage
    console.log('x.value = ' +
      analogVoltage);
    distanceInches = analogVoltage /
      0.00699;
    console.log("There is an object " +
    parseFloat(distanceInches).
      toFixed(3) + " inches away.");
}
```

Once this code is run, it will output the distance of the object to the console. Try moving the object closer or farther away; you will see that the distance updates every two seconds. If you want a more real-time update, simply decrease the time in the loop function; this will update the object's distance more frequently.

Project 9
Vibration Detection

This project utilizes the functionality of a piezoelectric transducer sensor, more commonly known as a piezo buzzer. A piezoelectric sensor detects pressure changes or acceleration by converting a force into an electrical current. For this project, we are going to detect a vibration or a knock on a door, and once we have detected this, we will light up an LED to indicate that someone is knocking at the door. This project is great for people with hearing disabilities because it alerts them when someone has knocked on their front door. This can also be used to alert you when someone is trying to break into a window.

Hardware

Table 3-2 lists the components and equipment required for Project 9.

TABLE 3-2	Components and Equipment for Project 9	
Schematic Reference	**Description**	**Appendix A**
	BeagleBone Black	M1
D1	5mm green LED	S5
R1	220Ω resistor	R1
	Piezo transducer	H5
	Jumper wires	H1
	Solderless breadboard	H2

The hardware used in this project consists of the component for a single green LED, as in Chapter 2, and a piezoelectric transducer with two wires connected to the solderless breadboard, as shown in Figure 3-5.

Software

Here is the code we'll use for Project 9:

```
var b = require('bonescript');

var led = "P8_10";

b.pinMode(led, b.OUTPUT);
```

Figure 3-5   Project 9 vibration detector

```
setInterval(check,500);

function check(){
b.analogRead('P9_40', checkKnock);

}

function checkKnock(x) {
    console.log (x.value);
    if (x.value <0.50) {
        b.digitalWrite(led, b.HIGH);
    }
    else
        b.digitalWrite(led, b.LOW);
  }
```

Initially, we set up the variables for the LED and define this as an output pin because we want the LED to light up when a knock is detected:

```
var b = require('bonescript');

var led = "P8_10";

b.pinMode(led, b.OUTPUT);
```

We now want to set a timing interval so we can check every so often for a knock at the door. We don't want this interval to be too high just in case someone knocks when our program isn't checking the analog input. Around 250ms should be enough (you can always play around with the timing). Once the function check has run, we simply use analogRead to read the status of the analog pin and then output the value to a function:

```
function check(){
b.analogRead('P9_40', checkKnock);
}
```

In this instance, the piezo transducer is connected to the analog pin of the BeagleBone Black, so we have a maximum voltage output of 1.8V and a minimum of 0V. This, in turn, gets converted to a digital value from 0 to 1. When the piezo transducer is in a standard operation state, the value is very low because there is no pressure on the sensor; therefore, it does not

generate as much voltage (only about 1V). When we push the sensor down, this generates a higher voltage. We can see this in our console output screen when it peaks at 0.99 just as the sensor is touched.

> **NOTE** Piezo sensors can be very sensitive to vibration, so do not press down too firmly.

To turn the LED on and off, we just create a simple if statement. This way, if the output value of the piezo sensor is greater than 0.8, we know there has been a knock at the door:

```
function checkKnock(x) {
    console.log (x.value);
    if (x.value >0.80) {
        b.digitalWrite(led, b.HIGH);
    }
    else
        b.digitalWrite(led, b.LOW);
```

This will turn on the LED (otherwise, the LED will be in an off state).

Project 10
GPS Tracker

This project will enable you to create your very own global positioning system using the BeagleBone Black's serial interface. This serial interface differs from the I2C bus in that it only allows for two-way communication between devices. The BeagleBone Black comes with up to five serial interface pins, and each serial port has a Tx (Transmit) and Rx (Receive) pin. These pins operate using the standard 3.3V, so before connecting your device to the BeagleBone Black, make sure it operates using this voltage as well.

Hardware

For this project we are going to keep the hardware very simple so that we can focus more on the

TABLE 3-3	Components and Equipment for Project 10
Description	**Appendix A**
BeagleBone Black	M1
Solderless breadboard	H2
Jumper wires	H1
Adafruit GPS module	M3

software side of things. The components and equipment for Project 10 are listed in Table 3-3.

We will use a GPS breakout board based on the popular MTK3339 module—it has everything we need to start tracking our location (see Figure 3-6).

The breakout board is built around the MTK3339 chipset. It is a high-quality GPS module that can track up to 22 satellites on 66 channels and has an excellent high-sensitivity receiver (–165 dB tracking) and a built-in antenna. It can perform up to 10 location updates a second for high-speed sensitivity logging or tracking. Power usage is incredibly low (only 20 mA during navigation), and an ultra-low dropout 3.3V regulator enables you to

power it with 3.3–5VDC in. The LED blinks at about 1 Hz while it's searching for satellites and blinks once every 15 seconds when a fix is found to conserve power. If you want to have an LED on all the time, you can use a FIX signal out on a pin to place an external LED.

This GPS breakout board only requires three jumper wires to the BeagleBone Black. See Figure 3-7 for the breadboard layout. You can see that we only require a 3.3V input to the breakout board and GND, but we only require one wire from the GPS module because we only need to receive the GPS coordinates. Therefore, we use the Tx pin on the GPS module and wire this straight into the Rx pin on the BeagleBone Black (see Figure 3-8). Make sure the voltage in the GPS sensor is connected to the 3.3V pin on the BeagleBone Black.

Software

Before we start writing our program, we need to install the serial communications manager on the BeagleBone Black because we are going to be using the serial pins for communication. Make sure you connect your BeagleBone Black to your Internet-enabled router/modem using an Ethernet cable and then open a Terminal window by navigating to the Bone101 web page and clicking GateOneSSH (this will open up a Terminal screen in your web browser). Next, type the following on the command line:

```
npm install -g serialport
```

This installs the serial library, where we can call the functions in our code and access all the serial pins on the BeagleBone Black. Here is the sketch code for Project 10:

```
var sp = require("serialport");

var port = '/dev/ttyO4';
var options = { baudrate: 9600, parser:
  sp.parsers.readline("\n") };
```

Figure 3-6 Adafruit ultimate GPS breakout board

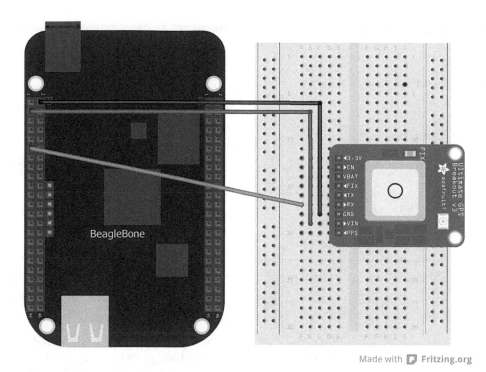

Figure 3-7 Project 10 breadboard layout

Figure 3-8 Project 10 GPS sensor

```
var gpsPort = new sp.SerialPort(port,
  options);

gpsPort.on('data', function(data) {
    //console.log('data received: ' +
      data);
    var parts = data.split(",");
    if (parts[0] == "$GPRMC") {
        var lat = parseFloat(parts[3])
          / 100;
        var ns = parts[4];
        var lon = parseFloat(parts[5])
          / 100;
        var ew = parts[6];
        console.log("Lat=" + lat + ns +
          " long=" + lon + ew);
    }
});
```

When you run the program, you will get latitude and longitude GPS coordinates on the console screen in the Cloud9 IDE, thus providing you have a fixed GPS signal. If you are unable to get a signal (which will be indicated on the GPS module's LED), you will receive the following message:

```
Lat=NaN long=NaN
```

In this case, you may want to move to an outdoor location, or by a window if you are inside. If you successfully get a signal, you should receive a message similar to the following (showing your GPS coordinates, of course):

```
Lat=53.463056N long=2.291389W
```

The first thing you may notice that is different in our code is that we are not referencing the "require bonescript" but instead are using serial port. This is because we do not need to use the BoneScript library to communicate with the GPIO pins; instead, we need the serial port library so we can communicate with the GPS module using the BeagleBone Black's serial pins:

```
var sp = require("serialport");
var port = '/dev/ttyO4';
```

The options variable is used to set up the serial communication options, such as the baud rate, which is the speed of communication between the GPS module and the BeagleBone Black. This value must match exactly the speed of the GPS module for communication. The parser option is used to specify how the data is read through the serial port. For our GPS module, we want to read the data one line at a time, which allows us to manage the incoming data better.

```
var options = { baudrate: 9600, parser:
  sp.parsers.readline("\n") };
```

We have now set up all our variables for serial communication, so the only thing left is to create the link between the two devices. To do this, we use the following code:

```
var gpsPort = new sp.SerialPort(port,
  options);
```

The code shown next defines how to handle the serial data coming in and how we can extract the GPS coordinates. The GPS module updates its position every second, and it sends data on a number of lines in different formats, such as the following:

```
$GPRMC,092610.000,A,5342.6381,N,00240.01
34,W,8.44,277.28,161213,,,A*7A
$GPVTG,277.28,T,,M,8.44,N,15.64,K,A*0B
$GPGGA,092611.000,5342.6380,N,00240.0155
,W,1,5,4.62,-38.2,M,49.0,M,,*55
$GPGSA,A,3,06,22,18,27,03,,,,,,,,4.72,
4.62,0.97*04
```

You can see that each line of data starts with $ and is followed by an identifier. These are known as NMEA sentence codes. The line we are most interested in is $GPRMC, which stands for global positioning recommended minimum specific transit data. If you take a look at Table 3-4, you can see the exact breakdown of the following line:

```
$GPRMC,092610.000,A,5342.6381,N,00240
.0134,W,8.44,277.28,161213,,,A*7A
```

TABLE 3-4	GPRMC Breakdown
Description	**GPS Data**
Time of fix	09:26:10 UTC
Receiver warning: A=ok, V=warning	A
Latitude	5342.6381N
Longitude	0240.0134W
Speed over ground (knots)	8.44
Course made good	277.28
Date of fix	16/12/13
Magnetic variation	-
Checksum	*7A

For this example, the data we need to extract is the latitude and the longitude values, and these are separated by commas and put into an array of strings. The handler function splits the data up into parts. Each part of the data we need we add to a variable so that we can organize the data how we wish.

```
gpsPort.on('data', function(data) {
    //console.log('data received: ' +
        data);
```

```
var parts = data.split(",");
if (parts[0] == "$GPRMC") {
    var lat = parseFloat(parts[3])
        / 100;
    var ns = parts[4];
    var lon = parseFloat(parts[5])
        / 100;
    var ew = parts[6];
```

We divide both the latitude and longitude values by 100 because they are 1/100ths of a degree. Once we have our variables for the GPS data, we can output the coordinates to the console:

```
console.log("Lat=" + lat + ns + " long="
    + lon + ew);
```

We should then see the following:

```
Lat=53.463056N long=2.291389W
```

To confirm that these coordinates are correct, we can directly input them to a mapping application such as Google Maps to pinpoint the exact GPS coordinates. See Figure 3-9 for the position of our current coordinates.

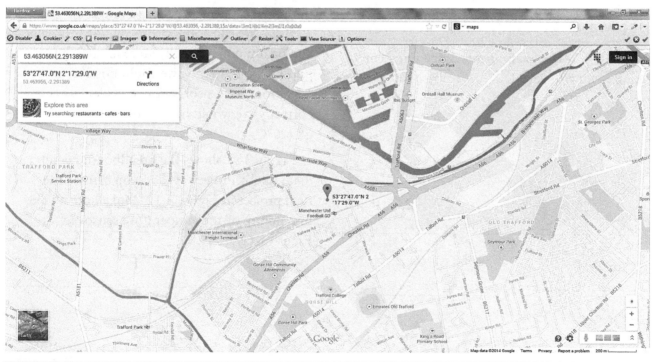

Figure 3-9 Google Maps GPS coordinates

To take this project a step further, you can output the coordinates to a text file, which can keep a log of where your GPS receiver has been. This is great for tracking weather balloons or rover buggies.

Project 11
Temperature Sensor

Probably one of the most popular sensors you will want to learn about is the temperature sensor, which enables you to read temperatures. There are many different ways in which this can be achieved, including using a thermistor across a voltage divider and measuring the output voltage across the resistors. We will create something similar to this in Project 13 using a light-dependent resistor. For this project, however, we will use a TMP36 sensor, shown in Figure 3-10. This is a simple device to use, and it cuts out the need for a resistive voltage divider because the output pin on the sensor outputs a voltage.

Hardware

Table 3-5 lists the components and equipment required for Project 11.

TABLE 3-5	Components and Equipment for Project 11	
Description	**Appendix A**
BeagleBone Black	M1
Solderless breadboard	H2
Jumper wires	H1
TMP36 temperature sensor	S6

NOTE Take extra caution when wiring the temperature sensor, as incorrect wiring may cause overheating and may burn the skin.

The TMP36 sensor has three pins, and it is very important that you get these pins oriented correctly when connecting the sensor to the breadboard. If in doubt, you can refer to the TMP36 technical datasheet, which is clearly illustrated. Pin 1 of the sensor is the voltage in from the BeagleBone Black, and this can be in the range of 2.7–5.5V, so we can connect our 3.3V from the BeagleBone. Pin 3 of the sensor is the GND, so we can connect this to any GND pin on the BeagleBone Black. Pin 2 is the important part of the sensor; this is the voltage out based on the temperature range of –40°C to +150°C. We can connect this pin directly to our analog input pin, which can read analog voltages up to 1.8V. In theory, we could possibly exceed 1.8V from the sensor, but if we calculate this correctly, we would have to exceed a temperature of 130°C.

Figure 3-11 shows the breadboard layout diagram for Project 11, and you can see the final project in Figure 3-12. Make note of the orientation of the temperature sensor in Figure 3-11.

Figure 3-10 TMP36 temperature sensor

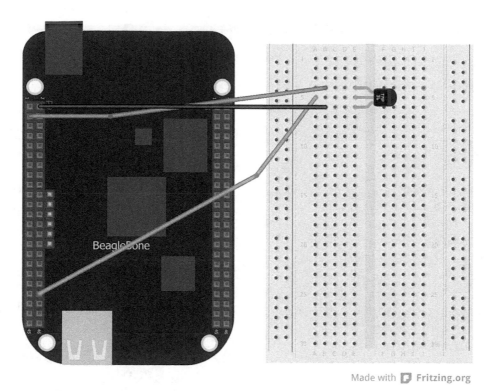

Figure 3-11 Project 11 breadboard layout

Figure 3-12 Project 11 temperature sensor

Software

Here is the code we'll use for Project 11:

```
var b = require('bonescript');

var inputPin = "P9_40";

getBeagleTemp();

function getBeagleTemp() {
    var value = b.analogRead(inputPin);

    var millivolts = value * 1800;
    var temp_c = (millivolts - 500)
        / 10;

    setTimeout(getBeagleTemp, 1000);

    console.log(temp_c);
}
```

The code for this project is really simple: all we are doing here is calculating the voltage coming in to the analog pin on the BeagleBone Black and converting this to degrees Celsius. The voltage in is converted into millivolts by multiplying the value by 1,800:

```
var millivolts = value * 1800;
```

We then convert the value into Celsius by subtracting 500 and then dividing the result by 10:

```
var temp_c = (millivolts - 500) / 10;
```

We could then use the following equation to convert from Celsius to Fahrenheit, if desired:

```
var temp_f = (temp_c * 9/5) + 32;
```

Another way to loop the function instead of setting an interval is to set a timeout in the function and just call the function again. We can set this code to update the temperature every second, like so:

```
setTimeout(getBeagleTemp, 1000);
```

Project 12
Moisture Sensor

Project 12 is for all you makers with a green thumb out there. This project is a fantastic DIY moisture level sensor that can detect when your vegetation needs watering. It uses a very simple circuit setup that measures the resistance between two probes. This allows you to sense whether the soil needs watering so you can keep your plants healthy.

Hardware

Table 3-6 lists the components and equipment required for Project 12.

TABLE 3-6	Components and Equipment for Project 12	
Schematic Reference	**Description**	**Appendix A**
	BeagleBone Black	M1
	Solderless breadboard	H2
	Jumper wires	H1
R1	10k resistor	R5
D1	Red LED	S1
D2	Green LED	S5
R2	220R resistor	R1

The hardware for this project is very simple: it uses a voltage divider to create a circuit where we can measure the resistance between two points. The two probes used for this project are two bare wires, but for convenience, you can solder them onto a pair of nails and stick them in the soil. This circuit diagram for Project 12 as shown in Figure 3-13 allows us to read the value of the sensor. When we need to water the vegetation or when the resistance of the soil increases, the green LED will light up indicating the vegetation needs watering. The final project can been seen in Figure 3-14.

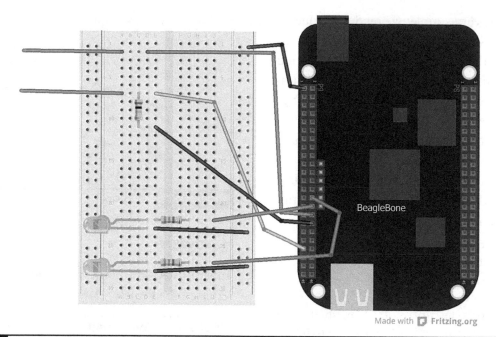

Figure 3-13 Breadboard layout for Project 12

Figure 3-14 Project 12 moisture sensor

Software

Here is the code we'll use for Project 12:

```
var b = require('bonescript');
var inputPin = "P9_40";

var ledPin = "P9_28";
var ledPin2 = "P9_30";

b.pinMode(ledPin, b.OUTPUT);
b.pinMode(ledPin2, b.OUTPUT);
loop();

function loop() {
    var value = b.analogRead(inputPin);
    console.log(value*1.8);
    if (value>0.5){
        b.digitalWrite(ledPin, b.HIGH);
        b.digitalWrite(ledPin2, b.LOW);
    }
    else b.digitalWrite(ledPin, b.LOW);
        b.digitalWrite(ledPin, b.HIGH);


    setTimeout(loop, 1000);
}
```

The code is very simple: we read the value between the two probes using the `analogRead` function and then determine whether this value is greater than 0.5 using an `if` statement. If it is, we know that the plant does not need watering; however, if the value is less than 0.5, we know there is greater resistance between the probes because of the moisture level. Therefore, it's time to water the plant. We also use two LEDs as a simple indicator: green for healthy moisture levels and red for poor moisture levels.

```
function loop() {
    var value = b.analogRead(inputPin);
    console.log(value*1.8);
    if (value>0.5){
        b.digitalWrite(ledPin, b.HIGH);
        b.digitalWrite(ledPin2, b.LOW);
    }
    else b.digitalWrite(ledPin, b.LOW);
        b.digitalWrite(ledPin, b.HIGH);
```

Project 13
Light Level Indicator

This project uses a light-dependent resistor (LDR) or photocell to detect levels of light (see Figure 3-15). The value changes depending on the amount of light received by the sensor: the resistance decreases as the light increases. This is a convenient and low-cost way of measuring basic light levels. This project cannot, however, be used for accurate measurement of lux levels, but it can tell us whether it is day or night. In the real world, this project provides a great way of turning on lights when its dark out (known as dusk-till-dawn lighting). Generally speaking, LDRs have a value of around 10k when it is dark; you may have to refer to the technical datasheet for specific information about your LDR.

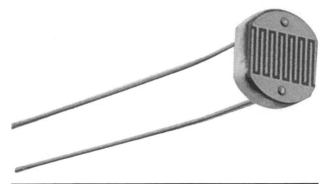

Figure 3-15   Photocell

Hardware

Table 3-7 lists the components and equipment required for Project 13.

TABLE 3-7	Components and Equipment for Project 13	
Schematic Reference	**Description**	**Appendix A**
	BeagleBone Black	M1
	Solderless breadboard	H2
	Jumper Wires	H1
R1	1k resistor	R8
R2	10k photocell	R7

As you know from our previous projects, the ADC on the BeagleBone Black can only read the voltage difference up to 1.8V. So, with this in mind, we need to create a divider circuit between the photocell and a pull-down resistor (see Figure 3-16). This allows us to measure the voltage drop across both the photocell and the resistor as the light levels change. For this project, we use a typical value of 1k for the pull-down resistor. Because we have connected our photocell to 1.8V ADC output on pin 32, we do not have to worry about the risk of damaging the board on pin 40 from a voltage higher than 1.8V or when the light resistance on the sensor has decreased.

You can see in Figure 3-16 that when there is no light, the circuit will be pulled to ground because the resistor's value of 1k would be

P9_40

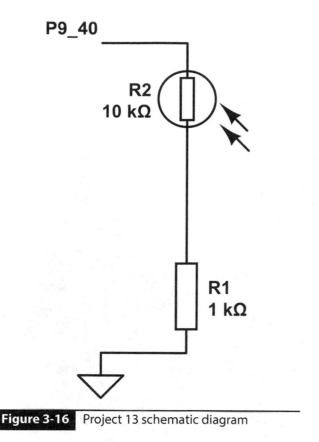

less than the assumed photocell value of 10k. When light starts to appear, the resistance of the photocell will drop and the voltage should be detected in pin P9_40.

Figure 3-17 shows the breadboard layout diagram for Project 13, and the final project can be seen in Figure 3-18.

Software

Here is the code we'll use for Project 13:

```
var b = require('bonescript');

var inputPin = "P9_40";

getlightlevel();

function getlightlevel() {
    var value = b.analogRead(inputPin);

    setTimeout(getlightlevel, 1000);

    console.log(value);
}
```

Figure 3-16   Project 13 schematic diagram

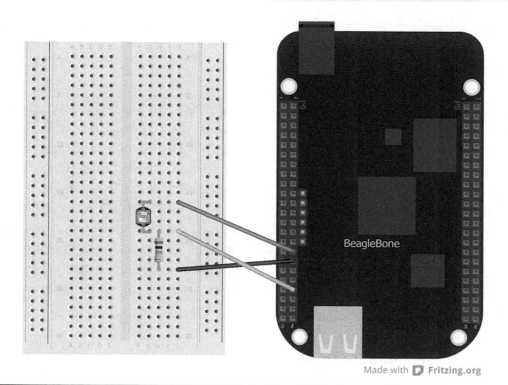

Made with **❚** Fritzing.org

Figure 3-17   Project 13 breadboard layout

Figure 3-18 Project 13 light level indicator

The code for this project is very simple: we create a function that updates every so often and reads the value from the analog pin. If we want to create a simple day/night indicator, we can hook up two LEDs so that when it is dark, the red LED will light up and when it is light, the green LED will light up. Refer back to Chapter 2 for setting up an LED.

a variety of analog sensors. When using analog sensors with the BeagleBone Black, remember that it is just a matter of calculating the voltage conversion you read from your analog pins. Hopefully, you can integrate these sensor projects with the LED projects in Chapter 2 to create some really great tools for sensing the real world around us. In the next chapter, we look at some robotics for controlling motors using a wireless keyboard and the Web.

Summary

Following the LED projects in the previous chapter, in this chapter, you learned how to use

CHAPTER 4

Robotic Projects

THIS CHAPTER LOOKS AT controlling servos and motors using the BeagleBone Black. These power devices can be used in applications such as robotics, CNC machinery, and automated systems. You have already read a bit about using pulse width modulation (PWM) in Chapter 2, and in this chapter, we will use the full functionality of PWM to control servos and motors as well as their speed. We will also be looking at controlling items through a web interface, which will enable our projects to be controlled remotely from anywhere in the world.

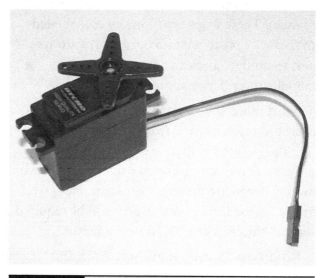

Figure 4-1 Standard servo motor

Project 14
Controlling Servos and Motors

Project 14 will give you a basic understanding of both servos and motors and how they are operated. By using the `analogWrite` function, the BeagleBone Black will send a PWM signal to the motor to control its operation.

Servo Motor Control

Although DC motors serve as excellent drive motors, they are not always best suited for precision because there is no way to tell the exact position of a DC motor. Servo motors are quite unique in the fact that you can control them to rotate to a particular position and they

will stay in that position unless told otherwise (see Figure 4-1).

Servo motors usually come in one of two types: full continuous rotation or standard rotation, which is usually 0 to 180 degrees because there is a potentiometer in line with the shaft for calculating the exact position. For this project, we will be using the standard (0 to 180 degrees) servo motor so that we can calculate its exact position.

Hardware

Table 4-1 shows the components and equipment required for Project 14a.

TABLE 4-1	Components and Equipment for Project 14a	
Schematic Reference	**Description**	**Appendix A**
	BeagleBone Black	M1
M1	5V servo motor	H6
	Breadboard	H2
	Jumper wires	H1
R1	1K resistor	R8

Unlike DC motors, servos have three pins: power (red), ground (black), and signal (white/orange). These wires are typically color coded so that it's easy to wire up devices. It's always best to double-check the datasheet of the manufacturer if you are unsure.

Most motors, when powered, can at times draw too much current from the power source (the BeagleBone in this case), but for this project, we are only going to drive one 5V servo motor. If you are driving more than one servo motor, an external power source will be required, such as batteries or a DC power adaptor.

Servos have a dedicated control pin that instructs them which position to turn the motor

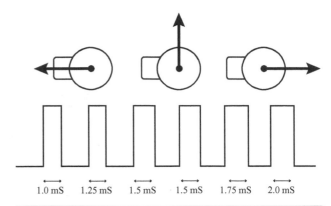

1.0 mS   1.25 mS   1.5 mS   1.5 mS   1.75 mS   2.0 mS

Figure 4-2   Servo motor timing diagram

to based on the adjustable PWM signals coming in. As shown in Figure 4-2, if we send a 1ms 5V pulse to the servo, it would turn the motor to 0 degrees (min), and if we send a 2ms 5V signal, it would turn the motor to 180 degrees (max). Using these calculations, we can quite easily work out that every 0.0111ms will turn the servo by 1 degree.

With all this in mind, we can now go ahead and wire up our servo motor to the BeagleBone Black, as shown in Figure 4-3. The completed project is shown in Figure 4-4.

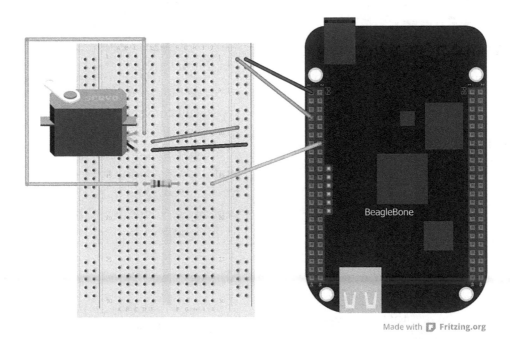

Made with [F] Fritzing.org

Figure 4-3   Servo motor breadboard layout

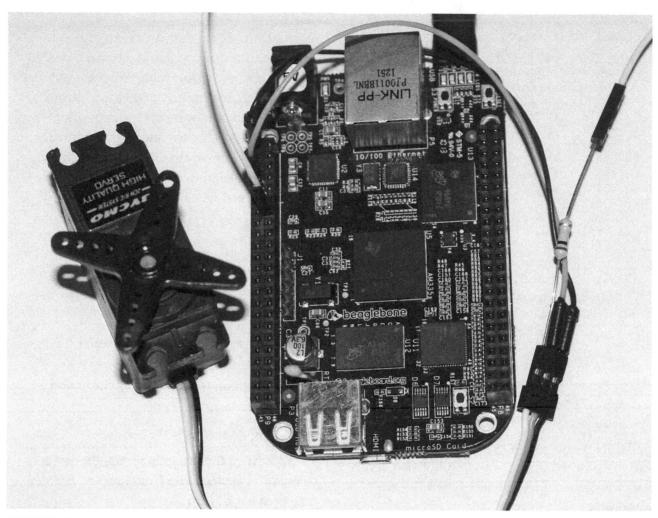

Figure 4-4 Controlling a servo motor

Software

The software for controlling a servo motor could not be any simpler. All we need to do is set

```
analogWrite(pin, value, freq, callback)
```

where *pin* is the PWM pin number we have our servo motor connected to (in this case, it is connected to P9_14), *value* is the position we want our servo to move to (this value would be our duty cycle), and *frequency* is the rate at which the PWM signal is sent to the servo motor (in Hz).

```
var b = require('bonescript');
var SERVO = 'P9_14';
var duty_min = 0.03;
var position = 0;
var increment = 0.1;

b.pinMode(SERVO, b.OUTPUT);
updateDuty();

function updateDuty() {
    // compute and adjust duty_cycle
    // based on desired position in
    // range 0..1
    var duty_cycle = (position*0.115)
        + duty_min;
```

```
b.analogWrite(SERVO, duty_cycle,
    60, scheduleNextUpdate);
    console.log("Duty Cycle: " +
        parseFloat(duty_cycle*100).
            toFixed(1) + " %");
}

function scheduleNextUpdate() {
    // adjust position by increment and
    // reverse if it exceeds range of
    // 0..1
    position = position + increment;
    if(position < 0) {
        position = 0;
        increment = -increment;
    } else if(position > 1) {
        position = 1;
        increment = -increment;
    }

    // call updateDuty after 200ms
    setTimeout(updateDuty, 200);
}
```

Before we do anything, we need to set the variables for the duty, the position of the servo, and how we want to increment the value (or rotation) of the motor:

```
var b = require('bonescript');
var SERVO = 'P9_14';
var duty_min = 0.03;
var position = 0;
var increment = 0.1;
```

The first function we call calculates the duty cycle; then we write the value to the servo motor using the analogWrite function:

```
function updateDuty() {
    // compute and adjust duty_cycle
    // based on desired position in
    // range 0..1
    var duty_cycle = (position*0.115)
        + duty_min;
    b.analogWrite(SERVO, duty_cycle, 60,
        scheduleNextUpdate);
    console.log("Duty Cycle: " +
        parseFloat(duty_cycle*100).
            toFixed(1) + " %");
}
```

After the value has been sent, we can then issue a callback (scheduleNextUpdate) to increment the duty cycle if it has not exceeded the maximum rotation value. We use an if statement to determine whether the value has reached a minimum or maximum position, and then we set the increment to a positive or negative value so that when we next run the update, the motor will turn in the correct rotation direction.

```
function scheduleNextUpdate() {
    position = position + increment;
    if(position < 0) {
        position = 0;
        increment = -increment;
    } else if(position > 1) {
        position = 1;
        increment = -increment;
    }
}
```

This function is run every 200ms to update to the next position:

```
setTimeout(updateDuty, 200);
```

Now, when the program is run, the servo should rotate clockwise from 0 to 180 degrees and then back again.

DC Motor Control

DC motors differ from servo motors in that when a voltage is applied across one, the motor is driven continuously in one direction. DC motors only have two contact pins: plus voltage and ground. So how do we control the speed and direction? In order to control the speed of the motor, we adjust the voltage supplied to it. Therefore, we know how to control the speed of a motor—but what about the direction? If we connect a voltage to a motor, it rotates in one direction, and if we swap the pin over, the motor rotates in the other direction. Therefore, all we have to do is reverse the voltage, and the motor simply goes in the other direction. This is generally done using an H-bridge IC, such as an L293. DC motors generally require a higher current than the BeagleBone Black can handle, so for this setup we require an external power source for our motors, such as batteries or a DC adaptor. This way, we can still control the speed using our H-bridge.

Hardware

Table 4-2 shows the components and equipment required for Project 14b.

TABLE 4-2	Components and Equipment for Project 14b	
Schematic Reference	Description	Appendix A
	BeagleBone Black	M1
	Solderless breadboard	H2
	Jumper wires	H1
M1	DC motor	H7
S1	L293d H-bridge	S7
	Four AA batteries	

For this part of the project, we are going to use a geared 6V motor to drive a small robotic chassis. To drive the motor, we use a Half-H driver such as the L293 IC. Technically, the L293d can drive up to four motors in one direction, but, of course, we want our motor to be controlled in both directions, so we will use one set of pins for clockwise and another set for counterclockwise. Looking at the technical datasheet of the L293, we can see that there are 16 pins on the IC (refer to Figure 4-5). A half circle at one end of the chip always indicates the top, and usually you can find a small circle next to pin 1.

Let's go through the pins just to get a better understanding of how they will be used:

- **GND (pins 4, 5, 12, and 13)** The four pins in the middle are connected to a shared ground.

- **Vcc2 (pin 8)** This pin supplies the motor current, so you connect this to your external power supply.

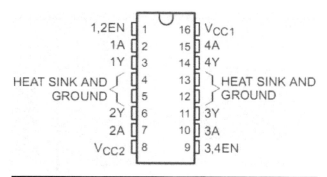

Figure 4-5 L293d H-bridge pinout

- **Vcc1 (pin 16)** This pin provides power for the chip, so we can connect this to the BeagleBone's 5V.

- **1Y and 2Y (pins 3 and 6)** These are the outputs from the left-side driver of motor 1. Our motor wires will connect directly to these pins.

- **1A and 2A (pins 2 and 7)** These pins control the state of the switches for motor 1. These are connected to the BeagleBone Black GPIO pins.

- **1, 2EN (pin 1)** This pin is used to enable or disable motor 1 driver. This pin will be connected to the BeagleBone Black's PWM pin to control the motor speed.

The rest of the pins are used to control motor 2, on the right side of the driver. The breadboard layout for Project 14b can seen in Figure 4-6, and the completed project is shown in Figure 4-7. You can see in Figure 4-7 that I have used a sliding potentiometer, but any standard potentiometer will do.

Software

Using a bit of logic, we can quite easily write our program to control the motor, both forward and backward:

```
var b = require('bonescript');
var pwmPin = "P9_14";
var ain1Pin = "P9_15";
var ain2Pin = "P9_17";
```

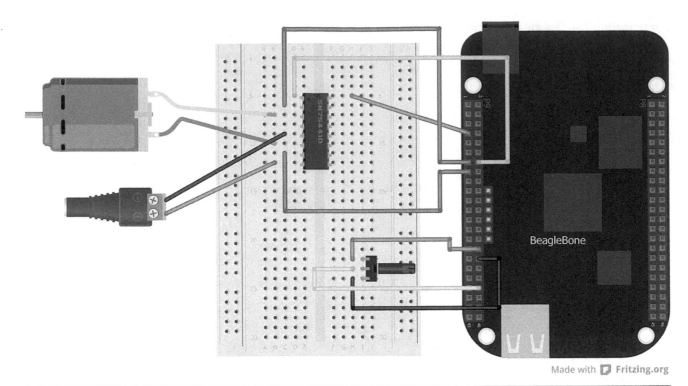

Made with **F** Fritzing.org

Figure 4-6 DC motor control breadboard layout

Figure 4-7 DC motor control using L293d

```
var potPin = "P9_33";

b.pinMode(pwmPin, b.OUTPUT);
b.pinMode(ain1Pin, b.OUTPUT);
b.pinMode(ain2Pin, b.OUTPUT);

function loop() {
    b.analogRead(potPin, adjustMotor);
}

function adjustMotor(reading) {
    if (reading.value > 0.5) {
        forwards((reading.value - 0.5)
            * 2);
    }
    else {
        backwards((0.5 - reading.value)
            * 2);
    }
}

function forwards(duty) {
    b.analogWrite(pwmPin, duty);
    b.digitalWrite(ain1Pin, 1);
    b.digitalWrite(ain2Pin, 0);
}

function backwards(duty) {
    b.analogWrite(pwmPin, duty);
    b.digitalWrite(ain1Pin, 0);
    b.digitalWrite(ain2Pin, 1);
}

setInterval(loop, 50);
```

The first part of the code sets up the variables and the pin state. We are using both analog and digital pins to control the motor using a POT.

```
var b = require('bonescript');
var pwmPin = "P9_14";
var ain1Pin = "P9_15";
var ain2Pin = "P9_17";
var potPin = "P9_33";

b.pinMode(pwmPin, b.OUTPUT);
b.pinMode(ain1Pin, b.OUTPUT);
b.pinMode(ain2Pin, b.OUTPUT);
```

After we set the variable, the next logical step is to get the value of the POT so we can then decide whether to turn the motor on forward or backward and at what speed. We read the value in a function so we can loop it to get a steady update of the status of the POT:

```
function loop() {
    b.analogRead(potPin, adjustMotor);
}
```

We can now calculate the direction using an `if` statement. If the trim pot value is greater than 0.5, we want to go forward, so we call the forwards function. If the value is less than 0.5, we want to go backward, so we call the backwards function.

```
function adjustMotor(reading) {
    if (reading.value > 0.5) {
        forwards((reading.value - 0.5)
            * 2);
    }
    else {
        backwards((0.5 - reading.value)
            * 2);
    }
}
```

Table 4-3 shows the logic in turning the pins high and low and how this drives the motor forward and backward.

TABLE 4-3	Motor Logic	
Status	**Ain1**	**Ain2**
Open	Low	Low
Forward	High	Low
Backwards	Low	High
Braking	High	High

When all the switches are in the open state, the motor will not spin. In the forward state, the first Ain pin is set to high and the other low, thus creating a circuit for the motor to spin forward. In the backward state, this is reversed so that the second Ain pin is set high and the first low, thus creating a reversing circuit and resulting in backward motor spin. When both Ain pins are high, this creates a braking state, causing the motor to stop its spin.

If we want the motor to run forward, we first set the PWM pin with the speed and then we set the Ain pins according to Table 4-3:

```
function forwards(duty) {
    b.analogWrite(pwmPin, duty);
    b.digitalWrite(ain1Pin, 1);
    b.digitalWrite(ain2Pin, 0);
}
```

To run the motor backward, we simply change the values of the pins to the opposite:

```
function backwards(duty) {
    b.analogWrite(pwmPin, duty);
    b.digitalWrite(ain1Pin, 0);
    b.digitalWrite(ain2Pin, 1);
}
```

We set the program to run every 50ms by using an interval timing, as shown next. This way, we can get a more accurate update of the trim pot position, which allows us to get better and faster control of the motor.

```
setInterval(loop, 50);
```

Project 15
Wireless Keyboard-Controlled Rover

Now that we have established how to turn on the motors, it's a good time to discuss how to control these motors to drive the direction for a small rover bot. For this project we are going to use a wireless keyboard and mouse controller to capture keypresses and control our rover. The rover we'll use is very simple: it has two motors on either side of the chassis driving the rubber tracks. Included is a battery compartment for four AA batteries to power the motors.

JavaScript isn't the only programming language we can use on the BeagleBone Black; another popular language among hobbyists is Python. In some ways, the programming concepts remain the same, regardless of which language is being used: we still need to call our library to access the GPIO pins and we set the pins to inputs or outputs.

Before we get started, we need to install a few items. First, we need to make sure the BeagleBone Black is up to date with the latest kernel (version 3.8). If you have followed along in this book since Chapter 1, yours should be fairly up to date. If not, refer back to Chapter 1 and update to the latest software. Connect your BeagleBone Black through SSH, either using GateOne SSH or using an SSH program such as PuTTY or ZOC Terminal. At this stage, make sure you are connected to an Internet-enabled cable/modem because you will need to download and install some software. Once you are connected, make sure the time and date are set; otherwise, you will have issues grabbing the software from the repository. Now type in the following line:

```
/usr/bin/ntpdate -b -s -u pool.ntp.org
```

Next, type in these commands to install some Python tools and the Adafruit GPIO library:

```
Opkg update && opkg install python-pip
python-setuptools python smbus
Pip install Adafruit_BBIO
```

As a quick test to make sure everything has installed correctly, you can just type the following:

```
python -c "import Adafruit_BBIO.GPIO as
GPIO; print GPIO"
```

Hardware

Table 4-4 shows the components and equipment required for Project 15.

The hardware for this project is very easy to set up. The key components are the motors and the L293d motor driver chip. We use the second part of the H-bridge to drive the second motor, and we control this through a PWM pin on the BeagleBone Black.

TABLE 4-4	Components and Equipment for Project 15	
Schematic Reference	Description	Appendix A
	BeagleBone Black	M1
	Jumper wires	H1
	Solderless breadboard	H2
S1	L293DNE motor driver	S7
	Pololu robot chassis	H8
M1/2	Two 6V geared motors	H7
	Wireless compact keyboard	H9
	Four AA batteries	

The breadboard layout diagram can be seen in Figure 4-8, and the completed project can be seen in Figure 4-9.

Software

Here is the Python code for driving the robot using a wireless keyboard in Project 15. The code allows us to drive the robot forwards and backwards and also turn it left and right.

```python
import os
import sys
import termios
import tty
import Adafruit_BBIO.GPIO as GPIO
import time
import Adafruit_BBIO.PWM as PWM

GPIO.setup("P9_14", GPIO.OUT)
GPIO.setup("P8_19", GPIO.OUT)

GPIO.setup("P9_13", GPIO.OUT) #motor1
GPIO.setup("P9_15", GPIO.OUT)

GPIO.setup("P8_10", GPIO.OUT) #motor2
GPIO.setup("P8_12", GPIO.OUT)


def getKey():
    fd = sys.stdin.fileno()
    old = termios.tcgetattr(fd)
    new = termios.tcgetattr(fd)
    new[3] = new[3] & ~termios.ICANON & ~termios.ECHO
    new[6][termios.VMIN] = 1
```

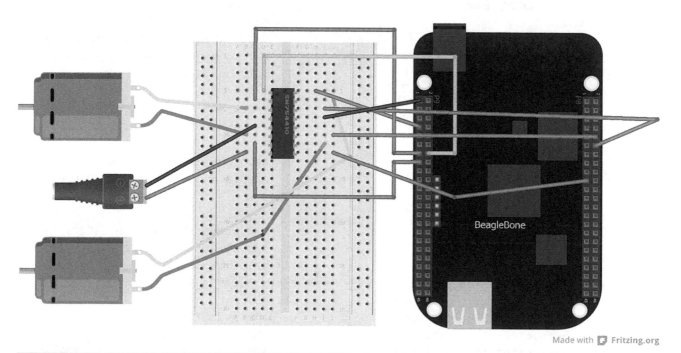

Figure 4-8 Breadboard layout for Project 15

Made with **Fritzing.org**

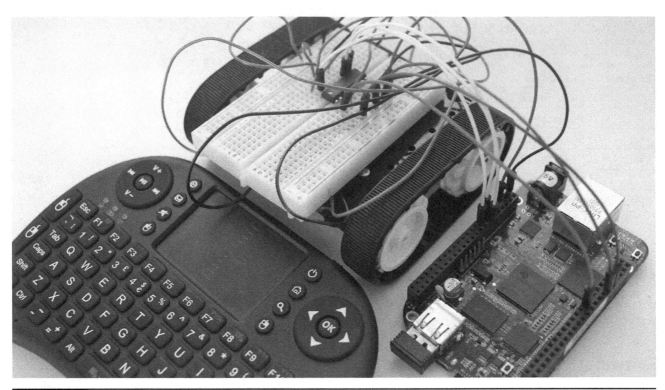

Figure 4-9 Project 15 wireless keyboard control rover

```
    new[6][termios.VTIME] = 0
    termios.tcsetattr(fd,
      termios.TCSANOW, new)
    key = None
    try:
        key = os.read(fd, 3)
    finally:
        termios.tcsetattr(fd,
          termios.TCSAFLUSH, old)
    return key


while 1:
   x = str(getKey())
   if x == "w":


      PWM.start("P9_14", 10, 2000, 1)
PWM.start("P8_19", 10, 2000, 1)
        GPIO.output("P9_13", GPIO.HIGH)
        GPIO.output("P9_15", GPIO.LOW)
        GPIO.output("P8_10", GPIO.HIGH)
        GPIO.output("P8_12", GPIO.LOW)
```

```
   elif x == "s":
      PWM.start("P9_14", 10, 2000, 1)
      PWM.start("P8_19", 10, 2000, 1)
      GPIO.output("P9_13", GPIO.LOW)
      GPIO.output("P9_15", GPIO.HIGH)
      GPIO.output("P8_10", GPIO.LOW)
      GPIO.output("P8_12", GPIO.HIGH)
   elif x == "a":
      PWM.start("P9_14", 10, 2000, 0)
      PWM.start("P8_19", 10, 2000, 1)

      GPIO.output("P8_10", GPIO.HIGH)
      GPIO.output("P8_12", GPIO.LOW)
   elif x == "d":

      PWM.start("P9_14", 10, 2000, 1)
      PWM.start("P8_19", 10, 2000, 0)
      GPIO.output("P9_13", GPIO.HIGH)
GPIO.output("P9_15", GPIO.LOW)
   else:
      PWM.stop("P9_14")
      PWM.stop("P8_19")
      PWM.cleanup()
```

Project 16
Web-Controlled Rover

In this project, we use a simple web interface to control a small rover using the BeagleBone Black. To achieve this, we use "web sockets" for the communication between the web interface and our node.js program. This project adopts a client/server model in which our BeagleBone JavaScript application creates a server that listens on a particular port number (in this case, 8080) and waits to receive the communication from the client. It then interfaces with the BeagleBone Black GPIO pins. The client side employs a simple HTML web page that has a slider as an input. This will be used to control the power and direction of the motors.

Hardware

Table 4-5 shows the components and equipment required for Project 16; many of these components were also used in Project 15.

For this project, we use a small robot chassis and two 6V geared motors. These items are relatively inexpensive and can be found in most hobbyist stores. Figure 4-10 shows the Pololu robot kit.

Figure 4-10 Pololu robot kit

TABLE 4-5	Components and Equipment for Project 16	
Schematic Reference	Description	Appendix A
	BeagleBone Black	M1
	Solderless breadboard	H2
	Jumper wires	H1
S1	L293d motor driver	S7
	Robot chassis	H8
M1/2	Two 6V DC motors	H7
	Four AA batteries	

As in Project 14, we will be using the L293d to drive two motors. This allows us to control the direction and speed of the rover. You can see the schematic diagram in Figure 4-11 and the completed project in Figure 4-12.

Software

The software for this project is split into two files: the application file (server) and the web HTML file (client). Before we can do anything, you must make sure you have the `socket.io` JavaScript program. To do this, type the following in the command line of the BeagleBone Black, but first make sure you are connected to the Internet using an Ethernet cable:

```
npm install socket.io
```

Once this has finished installing, we can create our server-side application:

```
var app = require('http').
createServer(handler);
var io = require('socket.io").
  listen(app);
var fs = require('fs');
var b = require('bonescript');

app.listen(8080);
// socket.io options go here
io.set('log level', 2);
// reduce logging - set 1 for warn,
// 2 for info, 3 for debug
```

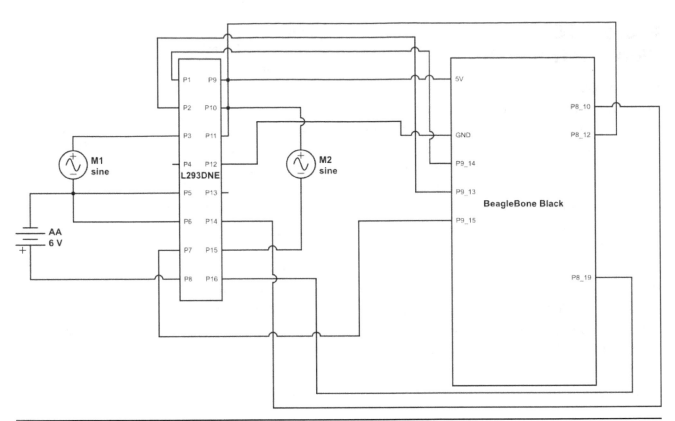

Figure 4-11 Schematic diagram for Project 16

Figure 4-12 Project 16 web-controlled rover

```
io.set('browser client minification',
  true);
// send minified client
io.set('browser client etag', true);
// apply etag caching logic based on
//   version number

console.log('Server running on:
  http://' + getIPAddress() + ':8080');
//pwm pin
var pwmPin1 = "P9_14";
var pwmPin2 = "P8_19";

//motor 1
var ain1Pin = "P9_13";
var ain2Pin = "P9_15";

//motor 2
var ain1Pin2 = "P8_10";
var ain2Pin2 = "P8_12";

// configure pins
b.pinMode(pwmPin1, b.OUTPUT);
b.pinMode(pwmPin2, b.OUTPUT);
b.pinMode(ain1Pin, b.OUTPUT);
b.pinMode(ain2Pin, b.OUTPUT);
b.pinMode(ain1Pin2, b.OUTPUT);
b.pinMode(ain2Pin2, b.OUTPUT);

function handler (req, res) {
  fs.readFile('index.html',
    // load html file
  function (err, data) {
    if (err) {
      res.writeHead(500);
      return res.end('Error loading
        index.html');
    }
    res.writeHead(200);
    res.end(data);
  });
}

io.sockets.on('connection', function
  (socket) {
  // listen to sockets and write analog
  // values to PWM pins
  socket.on('pwmPin', function (data) {
      console.log(data);
      if (data.value >50){
```

```
          forwards((data/100 - 0.5)
            * 2);
          console.log("forwards");
      }
      else {
          backwards ((0.5 - data/100)
            * 2);
          console.log("backwards");
      }
  });
});

function forwards(duty) {
    var value = duty;
  //Arm motors
  b.analogWrite(pwmPin1, value);
  b.analogWrite(pwmPin2, value);
  //Write values
  b.digitalWrite(ain1Pin, 1);
  b.digitalWrite(ain2Pin, 0);
  b.digitalWrite(ain1Pin2, 1);
  b.digitalWrite(ain2Pin2, 0);
}

function backwards(duty) {
    var value = duty;
    //Arm motors
    b.analogWrite(pwmPin1, value);
    b.analogWrite(pwmPin2, value);
    //Write values
    b.digitalWrite(ain1Pin, 0);
    b.digitalWrite(ain2Pin, 1);
    b.digitalWrite(ain1Pin2, 0);
    b.digitalWrite(ain2Pin2, 1);
}
// Get server IP address on LAN
function getIPAddress() {
  var interfaces = require('os').
    networkInterfaces();
  for (var devName in interfaces) {
    var iface = interfaces[devName];
    for (var i = 0; i < iface.length;
      i++) {
      var alias = iface[i];
      if (alias.family === 'IPv4' &&
        alias.address !== '127.0.0.1' &&
        !alias.internal)
        return alias.address;
    }
  }
```

```
    return '0.0.0.0';
}
```

In order to set up our server, we have to import the functions and JavaScript modules. First, we require `http` so we can create an HTTP server on the BeagleBone Black. Next, we require the `socket.io` client so we can communicate with the server using web sockets. Because we are parsing data, we need to be able to read the index.html file; therefore, we require file system (fs) features to do this.

```
var app = require('http').
  createServer(handler);
var io = require('socket.io').
  listen(app);
var fs = require('fs');
var b = require('bonescript');
```

In order to view the web page on the server, we set up a port number on which to listen for web client connections. Port 8080 is the standard web port number.

```
app.listen(8080);
```

Now that we have our variables set, we need to configure the `socket.io` settings using `io.set`. We set the log level to 2, which displays information about the socket connection and allows us to identify any errors easily. Because we are sending data over the Web, it's important to reduce the amount of traffic flowing through; therefore, we set the client minification to `true`, which eliminates all the noncode elements such as whitespace, thus reducing the size of the data being sent across. Another way to reduce the load on the web server is to create a caching option so that when the URL is requested again on the client side, the page is simply reloaded using an entity tag (etag) header.

```
io.set('log level', 2);
  // reduce logging - set 1 for warn,
  // 2 for info, 3 for debug
io.set('browser client minification', true);
  // send minified client
```

```
io.set('browser client etag', true);
  // apply etag caching logic based
  // on version number

console.log('Server running on:
  http://' + getIPAddress() + ':8080');
```

Now that we have everything set up, the next logical step is to load index.html into a function handler:

```
function handler (req, res) {
  fs.readFile('index.html',
    // load html file
  function (err, data) {
    if (err) {
      res.writeHead(500);
      return res.end('Error loading
        index.html');
    }
    res.writeHead(200);
    res.end(data);
  });
}
```

Now that we have loaded our HTML file, we can listen to the sockets and perform our calculations with the values received to write the PWM analog value. We receive the data from the slider in the index.html web page, and we calculate whether we want the rover to go forward or backward by using an `if` statement. If the value is greater than 50, we want the rover to go forward, and we calculate the PWM value to write to determine the speed of the motors.

```
io.sockets.on('connection', function
  (socket) {
  // listen to sockets and write analog
  // values to PWM pins
  socket.on('pwmPin', function (data) {
    console.log(data);
    if (data.value >50){
        forwards((data/100 - 0.5)
          * 2);
        console.log("forwards");
    }
    else {
        backwards ((0.5 - data/100)
          * 2);
```

```
                    console.log("backwards");
                }
        });
});
```

In Project 14, we created an H-bridge circuit to control whether the motor goes forward or backward. In this project, we set the GPIO pins on both motors to either HIGH or LOW to control the flow through the circuit. Our slider position determines which function to call, and then we write both the analog and digital values to drive the motor.

```
function forwards(duty) {
    var value = duty;
    //Arm motors
    b.analogWrite(pwmPin1, value);
    b.analogWrite(pwmPin2, value);
    //Write values
    b.digitalWrite(ain1Pin, 1);
    b.digitalWrite(ain2Pin, 0);
    b.digitalWrite(ain1Pin2, 1);
    b.digitalWrite(ain2Pin2, 0);
}

function backwards(duty) {
    var value = duty;
    //Arm motors
    b.analogWrite(pwmPin1, value);
    b.analogWrite(pwmPin2, value);
    //Write values
    b.digitalWrite(ain1Pin, 0);
    b.digitalWrite(ain2Pin, 1);
    b.digitalWrite(ain1Pin2, 0);
    b.digitalWrite(ain2Pin2, 1);
}
```

Now that we have our server script, we need to create an index.html file for our web interface:

```
<!DOCTYPE html>
<html>
<head>
    <meta charset="utf-8">
    <title>BeagleBone Black Demo</title>
    <script src="/socket.io/socket.
     io.js"></script>
    <script>
    var socket = io.connect();
    // Send data through socket
```

```
    function pwmPin(value){
        socket.emit('pwmPin', value);
    }
    </script>
</head>
<body>
<div data-role="page" id="page1">
    <div data-theme="a" data-role=
      "header">
        <h3>
            BeagleBone Black
        </h3>
    </div>
    <div data-role="content">
        <div data-role="fieldcontain">
            <label for="slider1">
                Motor Speed
            </label>
            <input id="slider1"
              type="range" name="slider"
              value="50"
             min="0" max="100"
            data-highlight="false"
                data-theme="b"
                onChange="pwmPin(value);">
        </div>
    </div>
</div>
</body>
</html>
```

The HTML here isn't very detailed because all we want is a simple web page with a slider. If you want to go into more detail using CSS styling and such, you can visit www.w3cschools .com for further information on using HTML and CSS. All we do here is connect to the server by creating a simple script that pushes the value of the slider to the server function:

```
<script src="/socket.io/socket.io.js">
  </script>
  <script>
  var socket = io.connect();
  // Send data through socket
  function pwmPin(value){
    socket.emit('pwmPin', value);
  }
  </script>
```

Figure 4-13 Project 16's web page

To run the program, first run the server script in the Cloud9 IDE. You will then see the IP address in the console. You can put the IP address in your browser to load the web page. Therefore, open up your browser and insert the IP address, followed by the port number we set up earlier (8080). Figure 4-13 shows this project's web page.

Project 17
Plant Hydration System

In Project 17, we will use a combination of Project 12 and Project 14 to detect the moisture levels in a plant pot and then switch on a water pump if the plant needs watering. This project is an example of how you can utilize what you have learned so far to create your own projects.

Hardware

Table 4-6 shows the components and equipment required for Project 17.

For this project, we are going to be using a 12V peristaltic pump to send water into the plant

TABLE 4-6	Components and Equipment for Project 17	
Schematic Reference	**Description**	**Appendix A**
	BeagleBone Black	M1
	Solderless breadboard	H2
	Jumper wires	H1
R1	10K resistor	R5
M1	12V peristaltic motor pump	H10
	9V battery PP3	
	Battery clip	
S1	L293d motor driver	H7

pot through a small PVC tube. Peristaltic pumps are more commonly used in medical professions for transporting clean sterile liquids. This pump works by rotating small ball bearings to squeeze the liquid through the PVC tube. Although this does not provide a lot of pumping pressure, it is ideal for watering a plant pot. To detect the moisture levels in the plant pot, we will be using the same hardware as in Project 12. Figure 4-14 shows the breadboard layout diagram for Project 17, and the completed project can be seen in Figure 4-15.

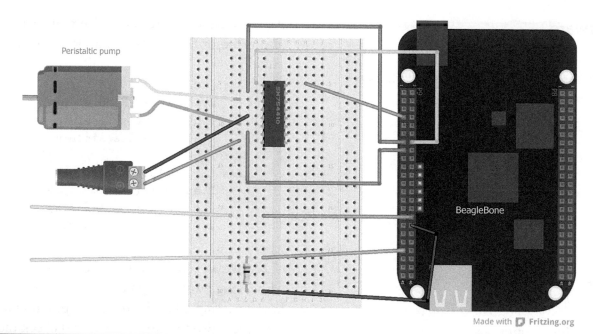

Peristaltic pump

BeagleBone

Made with ⬡ Fritzing.org

Figure 4-14 Project 17 breadboard layout diagram

Figure 4-15 Project 17 automatic watering system

Software

Here is the program code for Project 17. This program is very simple because it utilizes the code from Project 12 and Project 14.

```
var b = require('bonescript');

var inputPin = "P9_40";
var pwmPin = "P9_14";
var ain1Pin = "P8_10";
var ain2Pin = "P8_12";

b.pinMode(pwmPin, b.OUTPUT);
b.pinMode(ain1Pin, b.OUTPUT);
b.pinMode(ain2Pin, b.OUTPUT);
loop();

function loop() {
    var value = b.analogRead(inputPin);
    console.log(value);
    if (value<0.5){
        b.analogWrite(pwmPin, b.HIGH);
        b.digitalWrite(ain1Pin, 1);
        b.digitalWrite(ain2Pin, 0);
    }
    else if (value>0.5){
        b.analogWrite(pwmPin, 0);
        b.digitalWrite(ain1Pin, 0);
        b.digitalWrite(ain2Pin, 0);
    }

    setTimeout(loop, 1000);
}
```

The first thing we need to know is the value of the moisture sensor so we can determine whether or not to turn on the motor pump:

```
function loop() {
    var value = b.analogRead(inputPin);
```

Once we have gathered the value of the moisture sensor, we can then determine the output (TRUE or FALSE). If the value is less than 0.5, there is greater resistance between the two probes, which means the moisture levels are low and we need to water the plant pot. We then set the motor variable to turn on and water the plant. However, if the value is greater than 0.5, there is less resistance between the probes and

the moisture levels are good. Therefore, we turn off the motor because the plant does not need watering.

```
if (value<0.5){
        b.analogWrite(pwmPin, b.HIGH);
        b.digitalWrite(ain1Pin, 1);
        b.digitalWrite(ain2Pin, 0);
    }
    else if (value>0.5){
        b.analogWrite(pwmPin, 0);
        b.digitalWrite(ain1Pin, 0);
        b.digitalWrite(ain2Pin, 0);
    }
```

It can take some time to fill up the plant pot with enough water to get the correct moisture level, so it might be wise to set the program to loop every two to three minutes and check the levels again.

Project 18
Sentinel Turret

Project 18 is great for the office or workplace because you can acquire a target (such as a colleague) with your sentinel turret and then fire at will. This project is a great example of how you can use everyday USB devices and control them using the BeagleBone Black through the USB modules.

Hardware

Table 4-7 shows the components and equipment required for Project 18.

TABLE 4-7	Components and Equipment for Project 18	
Description		**Appendix A**
BeagleBone Black		M1
Dream Cheeky USB Thunder Missile Launcher		H11

This project uses the Dream Cheeky USB Thunder Missile Launcher, shown in Figure 4-16. It can rotate sideways, up, and down and

Figure 4-16 Project 18 USB sentinel turret

fire up to four missiles at a target. Best of all, it can be programmed through USB, so no soldering or external circuits are needed to control it.

Software

In order to control the missile launcher through USB, we need to download a few prerequisites. The first piece of software required is PyUSB, which is a library that allows us to access the USB port using the Python language. Run the following command in the Terminal window:

```
git clone git://github.com/walac/
  pyusb.git
python setup.py install
```

This will set up and install the PyUSB module. The second piece of software is a C library that gives applications such as PyUSB access to USB devices. To install Libusb, run the following command in the Terminal window.

```
Angstrom  Opkg install libusb-1.0-dev

Debian  sudo apt-get install
  libusb-1.0.0-dev
```

```python
import sys
import platform
import time
import urllib2
import usb.core
import usb.util

COMMAND_SETS = {
    "attack" : (
        ("zero", 0),
        ("led", 1),
        ("right", 3250),
        ("up", 540),
        ("fire", 4),
        ("led", 0),
        ("zero", 0),
    ),
}
DOWN    = 0x01
UP      = 0x02
LEFT    = 0x04
RIGHT   = 0x08
FIRE    = 0x10
STOP    = 0x20

DEVICE = None
DEVICE_TYPE = None
def usage():
    print "Usage: missile.py [command]
    [value]"
def setup_usb():
    global DEVICE
    global DEVICE_TYPE

    DEVICE = usb.core.
      find(idVendor=0x2123,
      idProduct=0x1010)

    if DEVICE is None:
        DEVICE = usb.core.
          find(idVendor=0x0a81,
          idProduct=0x0701)
        if DEVICE is None:
            raise ValueError('Missile
              device not found')
        else:
            DEVICE_TYPE = "Original"
    else:
        DEVICE_TYPE = "Thunder"
```

```python
    if "Linux" == platform.system():
        try:
            DEVICE.detach_kernel_
                driver(0)
        except Exception, e:
            pass

    DEVICE.set_configuration()

def send_cmd(cmd):
    if "Thunder" == DEVICE_TYPE:
        DEVICE.ctrl_transfer(0x21, 0x09,
            0, 0, [0x02, cmd,
        0x00,0x00,0x00,0x00,0x00,0x00])
    elif "Original" == DEVICE_TYPE:
        DEVICE.ctrl_transfer(0x21, 0x09,
            0x0200, 0, [cmd])

def led(cmd):
    if "Thunder" == DEVICE_TYPE:
        DEVICE.ctrl_transfer(0x21, 0x09,
            0, 0, [0x03, cmd,
        0x00,0x00,0x00,0x00,0x00,0x00])
    elif "Original" == DEVICE_TYPE:
        print("There is no LED on this
            device")
def send_move(cmd, duration_ms):
    send_cmd(cmd)
    time.sleep(duration_ms / 1000.0)
    send_cmd(STOP)

def run_command(command, value):
    command = command.lower()
    if command == "right":
        send_move(RIGHT, value)
    elif command == "left":
        send_move(LEFT, value)
    elif command == "up":
        send_move(UP, value)
    elif command == "down":
        send_move(DOWN, value)
    elif command == "zero" or command
        == "park" or command == "reset":
        # Move to bottom-left
        send_move(DOWN, 2000)
        send_move(LEFT, 8000)
    elif command == "pause" or command
        == "sleep":
        time.sleep(value / 1000.0)
    send_cmd(STOP)
```

```python
def run_command(command, value):
    command = command.lower()
    if command == "right":
        send_move(RIGHT, value)
    elif command == "left":
        send_move(LEFT, value)
    elif command == "up":
        send_move(UP, value)
    elif command == "down":
        send_move(DOWN, value)
    elif command == "zero" or command
        == "park" or command == "reset":
        # Move to bottom-left
        send_move(DOWN, 2000)
        send_move(LEFT, 8000)
    elif command == "pause" or command
        == "sleep":
        time.sleep(value / 1000.0)
    elif command == "led":
        if value == 0:
            led(0x00)
        else:
            led(0x01)
    elif command == "fire" or command
        == "shoot":
        if value < 1 or value > 4:
            value = 1
            time.sleep(0.5)
        for i in range(value):
            send_cmd(FIRE)
            time.sleep(4.5)
    else:
        print "Error: Unknown command:
            '%s'" % command
def run_command_set(commands):
    for cmd, value in commands:
        run_command(cmd, value)


def main(args):

    if len(args) < 2:
        usage()
        sys.exit(1)
    setup_usb()
    command = args[1]
    value = 0
    if len(args) > 2:
        value = int(args[2])
    if command in COMMAND_SETS:
```

```
        run_command_set(COMMAND_
            SETS[command])
    else:
        run_command(command, value)
def main(args):
    if len(args) < 2:
        usage()
        sys.exit(1)
    setup_usb()
    command = args[1]
    value = 0
    if len(args) > 2:
        value = int(args[2])
    if command in COMMAND_SETS:
        run_command_set(COMMAND_
            SETS[command])
    else:
        run_command(command, value)
if __name__ == '__main__':
    main(sys.argv)
```

To run the program and control the missile launcher, type the following command followed by two arguments—the command and value—as detailed in Table 4-8.

```
Sudo ./missile.py [command] [value]
```

Summary

By using what you have learned in this chapter, you can create some robotics projects and control them through the Web or a wireless keyboard. You can also implement what you learned in Chapter 3 to sense the environment to create a truly autonomous robot using basic artificial intelligence. The next chapter looks at displaying information using a variety of LED and LCD display devices.

TABLE 4-8   Missile Commands

Command	Value	Description
Up	Milliseconds	Moves up for the duration set
Down	Milliseconds	Moves down for the duration set
Left	Milliseconds	Moves left for the duration set
Right	Milliseconds	Moves right for the duration set
Fire	1–4	Fires the number of missiles indicated
Zero	Null	Sets the position to the bottom left
Pause	Null	Pauses between commands
Led	0–1	Turns the LED on or off
Command_set	Null	Programs the command set in the code and runs the command to perform a sequence

CHAPTER 5

Display Projects

THIS CHAPTER IS ALL ABOUT displaying data using a variety of devices, such as a four-digit LED display and a 16×2 character LCD display. If your project involves a lot of statistical information from sensors or other devices, it is always a good idea to display the information on a small display instead of outputting the information to the console or over SSH.

Project 19
7-Segment Clock

This project uses a four-digit 7-segment LED display to show the current system time. A 7-segment display is great for displaying decimal digits and is generally inexpensive. It's called a 7-segment display because—well, you guessed it—it has seven segmented LEDs on the interface. Figure 5-1 details how the display is labeled, from A to G.

Hardware

Table 5-1 lists the components and equipment required for Project 19. For this project we will use the Adafruit 7-segment display with an I2C backpack (see Figure 5-2). We could directly connect all 12 pins of the 7-segment display to the BeagleBone Black and control it that way, but this is not practical in the real world. Using

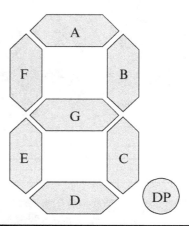

Figure 5-1 A 7-segment display

TABLE 5-1	Components and Equipment for Project 19	
Description		**Appendix A**
BeagleBone Black		M1
Solderless breadboard		H2
Jumper wires		H1
Adafruit 7-segment display		M5

an I2C chip reduces the number of jumper wires needed and simplifies our project somewhat.

The 7-segment display kit does require some basic-level soldering to attach the display to the I2C backpack. You must also solder four pin headers so that you can insert the display into your breadboard. Figure 5-3 shows the breadboard layout for this project.

Figure 5-2 Adafruit four-digit 7-segment display

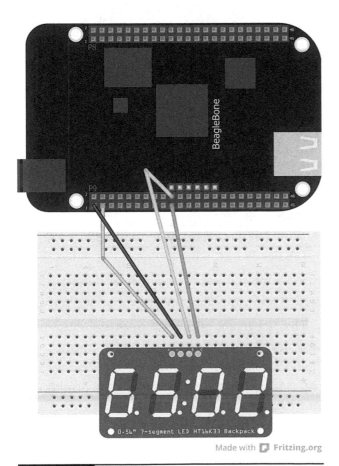

Figure 5-3 Breadboard layout for Project 1

Software

If you have not done so already, you need to install the Adafruit BeagleBone Black Python library, which allows you to control the BeagleBone Black's GPIO pins using Python. The first step is to log into the BeagleBone Black through SSH and set the current date and time by typing the following command:

```
ntpupdate -b -s -u pool.ntp.org
```

Next, install the Python packages and modules required:

```
opkg update && opkg install python-pip
python-setuptools python-smbus
pip install Adafruit_BBIO
```

You then need to download and install the Adafruit Python Raspberry Pi library using `github`. This library includes the Python modules that allow you to use the 7-segment display backpack.

```
git clone git://github.com/adafruit/
   Adafruit-Raspberry-Pi-Python-Code.git
```

Before you run the clock demo, it is best to check first whether you have the LED display connected properly, and that you can see it in the I2C table, by typing the following on the command line:

```
i2cdetect —y —r 1
```

You should see an I2C table, and the LED display should have an address of 0x70 so that you know it is connected to the BeagleBone and that you can communicate with the device. To run the program, "change directory" to the Adafruit LED backpack by typing the following command on the command line:

```
cd /Adafruit-Raspberry-Pi-Python-Code/
   Adafruit_LEDBackpack
```

Type ls, and you should see a list of all the files and folders in that current directory. In this directory you should see all the programs that run the LED backpack, such as our 7-segment display or an LED matrix board. To run the clock program that displays the current system time, type the following command:

```
python ex_Adafruit_Raspberrypi_clock
```

Figure 5-4 shows the Adafruit LED backpack clock in action.

Project 20
Displaying Sensor Information

LED displays are also great for showing values from sensors, such as a temperature sensor. The circuitry is very simple and consumes only minimal power, yet it provides a bright display for key information.

Hardware

Table 5-2 lists the components and equipment required for Project 20, and Figure 5-5 is the breadboard layout diagram for the circuitry. Figure 5-6 shows the completed project.

Figure 5-4 Project 19 Adafruit LED backpack clock

TABLE 4-2	Components and Equipment for Project 20
Description	**Appendix A**
BeagleBone Black	M1
Solderless breadboard	H2
Jumper wires	H1
TMP36 temperature sensor	S6
7-segment LED backpack	M5

For this project we are going to use the TMP36 temperature sensor we used in Chapter 3. This sensor is ideal for this project because it gives us a direct voltage output from the middle pin—essentially, it creates a voltage divider circuit for us.

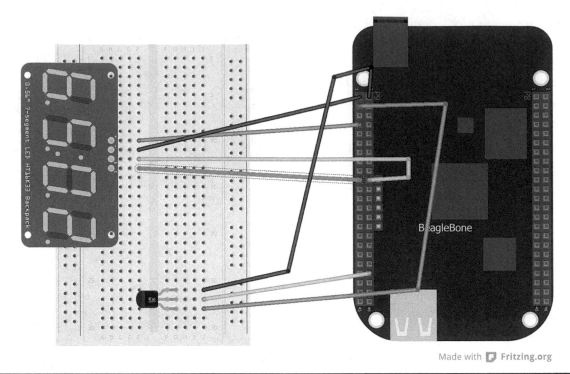

Figure 5-5 Breadboard layout for Project 20

Figure 5-6 Project 20 displaying temperature

Software

Here is the Python code we'll use for Project 20:

```python
import Adafruit_BBIO.ADC as ADC
import time
import datetime
from Adafruit_7Segment import
SevenSegment

segment = SevenSegment(address=0x70)
sensor_pin = 'P9_40'

ADC.setup()

while(True):
    reading = ADC.read(sensor_pin)
    millivolts = reading * 1800
    temp_c = (millivolts - 500) / 10

    segment.writeDigit(0, int(temp_c / 10))
    segment.writeDigit(1, int(temp_c % 10))
    segment.writeDigit(3, 12)

    segment.setColon(1)
    time.sleep(1)
```

The code is very simple: we are using the I2C Adafruit library, and we just need to read the value from the sensor, which will be in millivolts, and then convert that into a readable value such as degrees Celsius, as shown here:

```
reading = ADC.read(sensor_pin)
    millivolts = reading * 1800
    temp_c = (millivolts - 500) / 10
```

Once we have the temperature value, we write each digit on the 7-segment display using `segment.writeDigit(digit, value)`, where `digit` is the digit we want to write to in the four-digit display, from 0 to 3. Because we can only display one number per digit, we need to do some simple mathematic calculations so that we return the first digit of the temperature. To get the first digit, we simply divide the temperature value by 10, and this will be our first number to write to the four-digit display. Getting the second digit is little bit trickier: we can use the % symbol to divide the temperature value by 10. This will return the remainder value (that is, the second digit).

```
segment.writeDigit(0, int(temp_c / 10))
segment.writeDigit(1, int(temp_c % 10))
```

Now that we have our digits to display, we can write to the third digit on the display the letter *C,* for Celsius. The Adafruit library has these characters labeled from 0 to 15, which represent 0–9 and A–F. Therefore, the letter *C* would be the number 12.

```
segment.writeDigit(3, 12)
```

Project 21
LCD Display

One of the great things about designing and developing embedded systems is that they can be operated on their own. Up until this chapter, we have been plugging our BeagleBone Black into a computer and using the console or terminal to display the information we require. Now, by adding an LCD display to our projects, we can easily display more complex information such as sensor values, settings, progress information, and much more. In this project you will learn how to connect your LCD display to the BeagleBone Black and display some text on the screen using the BoneScript library.

Hardware

Table 5-3 lists the components and equipment required for Project 21. This project uses a parallel LCD screen, which is a common device that can be quite inexpensive. These devices come in an array of different shapes and sizes.

TABLE 5-3	Components and Equipment for Project 21	
Description	**Appendix A**
BeagleBone Black	M1
Solderless breadboard	H2
Jumper wires	H1
16×2 LCD screen	M6
10k potentiometer	R4

The most common size—and the one we will be using for this project—is a 16×2 character display with a single row of 16 pins, including a backlight LED for brightness. Most LCD displays do not come with a soldered 16-pin header on the printed circuit board (PCB), so you may want to solder one on so you can attach the display to your breadboard (see Figure 5-7).

Now you need to wire up the LCD display to the BeagleBone Black using Table 5-4 as a guide. All the parallel LCD displays have the same pinout, and they can be wired in one of two different modes: either four-pin mode or eight-pin mode. For this project, we will be using a four-pin wire configuration.

Figure 5-7 LCD display with headers soldered on

TABLE 5-4	Parallel LCD Pins	
Pin Number	**Pin Name**	**Description**
1	VSS	Ground
2	VDD	+5V
3	V0	Contrast (POT)
4	RS	Register
5	RW	Read/Write
6	EN	Enable
7	D0	Data 0
8	D1	Data 1
9	D2	Data 2
10	D3	Data 3
11	D4	Data 4
12	D5	Data 5
13	D6	Data 6
14	D7	Data 7
15	A	Backlight anode
16	K	Backlight cathode

The contrast pin changes the brightness of the LCD display using the value from an external potentiometer (usually a 10k potentiometer). The register pin uses the LCD to determine the next set of data to be written: either a command

is sent to move the cursor or a character is written to the screen. The RW pin is always set to 0 for this project, which means we always want to write data to the LCD display and never read the data on the screen. We use the EN pin to tell the LCD display when we are ready to send the data. Pins 4–7 on the LCD display are used to transmit the data to the screen. Because we are using four-pin mode, pins 0–3 are not connected.

The breadboard layout for Project 21 is shown in Figure 5-8. This can be quite a tricky process, so take care and make sure each wire is connected correctly.

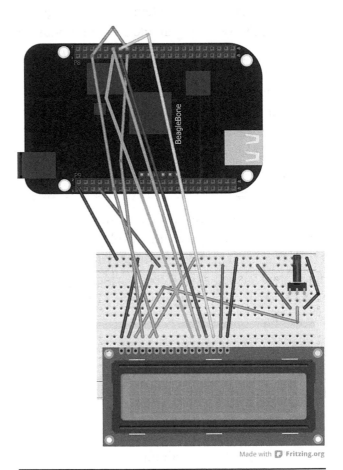

Made with [] Fritzing.org

Figure 5-8 Breadboard layout for Project 21

Software

Here is the JavaScript code we'll use for Project 21:

```javascript
var b = require('bonescript');

var my_string = "BeagleBone Black Evil Genius";

var lcd_pin_D4 = "P8_11";
var lcd_pin_D5 = "P8_12";
var lcd_pin_D6 = "P8_13";
var lcd_pin_D7 = "P8_14";
var lcd_rs = "P8_15";
var lcd_e = "P8_16";

b.pinMode(lcd_pin_D4, b.OUTPUT);
b.pinMode(lcd_pin_D5, b.OUTPUT);
b.pinMode(lcd_pin_D6, b.OUTPUT);
b.pinMode(lcd_pin_D7, b.OUTPUT);
b.pinMode(lcd_rs, b.OUTPUT);
b.pinMode(lcd_e, b.OUTPUT);

b.digitalWrite(lcd_pin_D4, b.LOW);
b.digitalWrite(lcd_pin_D5, b.LOW);
b.digitalWrite(lcd_pin_D6, b.LOW);
b.digitalWrite(lcd_pin_D7, b.LOW);
b.digitalWrite(lcd_rs, b.LOW);
b.digitalWrite(lcd_e, b.LOW);


LCD_init( function () { write_string_to_lcd(my_string, function() {} ); } );

function LCD_init(callback) {
    b.digitalWrite(lcd_e, b.LOW);
    var i = 0;
    var steps = [
        function(){ setTimeout(next, 50); },
        function(){ write4bits(0x03, next); },
        function(){ setTimeout(next, 5); },

        function(){ write4bits(0x03, next); },
        function(){ setTimeout(next, 5); },

        function(){ write4bits(0x03, next); },
        function(){ setTimeout(next, 2); },
        function(){ write4bits(0x02, next); },
        function(){ write_bits_to_lcd(0x28, true, next); },
        function(){ write_bits_to_lcd(0x0C, true, next); },
        function(){ write_bits_to_lcd(0x06, true, next); },
        function(){ write_bits_to_lcd(0x01, true, next); },
```

```
        function(){ write_string_to_lcd('\x7escreen init', next ); },
        function(){ setTimeout(next, 800); },
        function(){ write_bits_to_lcd(0x01, true, callback); }
        ];
    next();
    function next() {
        i++;
        steps[i-1]();
    }
}
function write_bits_to_lcd(value, command_or_character, callback) {
    var value_left = (value >> 4);
    if (command_or_character === true) b.digitalWrite(lcd_rs, b.LOW);
        else b.digitalWrite(lcd_rs, b.HIGH);
    write4bits(value_left, function() { write4bits(value, callback); } );
}
function write4bits(value, callback) {
    if((value >> 3) & 0x01) b.digitalWrite(lcd_pin_D7, b.HIGH);
        else b.digitalWrite(lcd_pin_D7, b.LOW);
    if((value >> 2) & 0x01) b.digitalWrite(lcd_pin_D6, b.HIGH);
        else b.digitalWrite(lcd_pin_D6, b.LOW);
    if((value >> 1) & 0x01) b.digitalWrite(lcd_pin_D5, b.HIGH);
        else b.digitalWrite(lcd_pin_D5, b.LOW);
    if(value & 0x01) b.digitalWrite(lcd_pin_D4, b.HIGH);
        else b.digitalWrite(lcd_pin_D4, b.LOW);
    b.digitalWrite(lcd_e, b.HIGH);
    b.digitalWrite(lcd_e, b.LOW);
    callback();
}
function write_string_to_lcd(s, callback) {
    var s_length = s.length;
    if (s_length > 32) {
        write_string_to_lcd_two_lines('ERROR:', 'String too long', callback);
        return;
    }
    if (s_length < 17) {
        write_bits_to_lcd(0x80, true, function () {
 helper_lcd_display_string(s, callback ); } );
    } else {
        write_string_to_lcd_two_lines(s.slice(0,16),
 s.slice(16, s_length), callback);
    }
}
function write_string_to_lcd_two_lines(line1, line2, callback) {
    write_bits_to_lcd(0x80, true, function () {
 helper_lcd_display_string(line1, function() {second_line();} ); } );
    function second_line() {
        write_bits_to_lcd(0xC0, true, function () {
 helper_lcd_display_string(line2, callback ); } );
```

```
    }
}
function helper_lcd_display_string(s, callback) {
    if (s.length === 0) {
        callback();
    } else {
        write_bits_to_lcd(s.charCodeAt(0), false, function()
            {helper_lcd_display_string(s.slice(1), callback);} );
    }
}
```

Project 22
LED Matrix Scrolling Text Display

This project uses the LED matrix display backpack; however, we are going to combine two of them to create a scrolling text display. Scrolling displays and message boards are used in a variety of embedded applications, such as to display travel information or stock/share values. These message displays are great for showing a lot of information in such a short amount of time but still getting the information across clearly. This project is based on LED Stock Ticker by Matt Hassel, which was written to display a variety of information across multiple matrix displays for the Raspberry Pi. I have adapted the code for use with the BeagleBone Black.

Hardware

Table 5-5 lists the components and equipment required for Project 22.

TABLE 5-5	Components and Equipment for Project 22	
Description	**Appendix A**
BeagleBone Black	M1
Solderless breadboard	H2
Jumper wires	H1
Adafruit 8×8 matrix display	M7

The first thing you need to do is assemble the LED matrix displays, if you haven't done so already, paying particular attention to the solder tabs on the back of the I2C backpack. By default, the matrix display comes with the I2C address of 0x70, and because you are adding two of these backpacks, you need to change the address on one of them so they both have a unique I2C address. This will also help you avoid any collisions that may occur. This can be done in two ways: either by soldering the pins together on the back of the I2C backpack or by using the software on the BeagleBone Black. For this project we are going to change the I2C address using the solder pins on the back of the LCD matrix. Figure 5-9 shows labels A0, A1, and A2; each one represents a value that gets added to the default address. Thus, A0 is equal to +1, A1 is equal to +2, and A2 is equal to +4. If you shorten the solder pin A0, your address would become 0x70 + 1 = 0x71.

Fortunately, both matrix displays use the same pins on the BeagleBone Black, so you only need to use four wires to connect (see Figure 5-10). You can see in Figure 5-11 that the second matrix display is connected to the adjacent pins on the first matrix display. This is due to the way the I2C bus works.

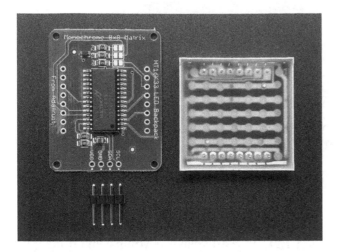

Figure 5-9 LED matrix I2C address solder pins

Software

Here is the Python code we'll use for Project 22:

```
#!/usr/bin/python
import time
import string2text as str2mat

app_text = str2mat.Text2LED()
app_ticker = str2mat.LED_TICKER()

message = time.strftime('%c')

while True:
        app_ticker.ticker =
            app_text.add_to_ticker(message
            + " ")
app_ticker.main()
```

This program is very simple. We first import the `string2text` library, which gets our message characters and converts them into a string format that can be written to the matrix displays:

```
import string2text as str2mat

app_text = str2mat.Text2LED()
app_ticker = str2mat.LED_TICKER()
```

We then create a variable called `message` that we can set as the time and date using the system time we imported:

```
message = time.strftime('%c')
```

We can display the time and date in a variety of formats using the values in Table 5-6.

Just remember that you can only display the characters in the alphabet string library; some characters, such as the colon and backslash, may not be able to be displayed unless you add them to the list of characters.

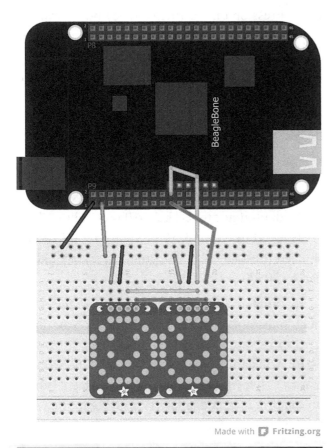

Made with **Fritzing.org**

Figure 5-10 Breadboard layout for Project 22

Figure 5-11 Project 22 scrolling message board

TABLE 5-6	Date Formats
Date Formats	Date Formats
%a	Weekday name
%A	Full weekday name
%b	Abbreviated month name
%B	Full month name
%c	Appropriate date and time
%d	Day of month as a decimal
%H	24-hour clock
%I	12-hour clock
%j	Day of the year as a decimal
%m	Month as a decimal
%M	Minute as decimal
%p	AM or PM
%s	Seconds as a decimal
%U	Weeks
%w	Weekday as a decimal
%W	Week number
%x	Appropriate date
%X	Appropriate time
%y	Year as a decimal
%Y	Year with century as a decimal
%z	Time zone
%%	% character

To display a letter of the alphabet, you must first get the string array for the matrix display. To do this, simply write a value of 1 or 0, where 1 is the LED in an "on" state and 0 is the LED in an "off" state:

```
myalphabet = { "A": [[0,1,1,0,0],
                     [1,0,0,1,0],
                     [1,0,0,1,0],
                     [1,1,1,1,0],
                     [1,0,0,1,0],
                     [1,0,0,1,0],

                     [1,0,0,1,0],
                     [0,0,0,0,0]],
              "B": [[1,1,1,0,0],
                     [1,0,0,1,0],
                     [1,0,0,1,0],
                     [1,1,1,0,0],
                     [1,0,0,1,0],
                     [1,0,0,1,0],
                     [1,1,1,0,0],
                     [0,0,0,0,0]],
              "C": [[0,1,1,1,0],
                     [1,0,0,0,0],
                     [1,0,0,0,0],
                     [1,0,0,0,0],
                     [1,0,0,0,0],
                     [1,0,0,0,0],
                     [0,1,1,1,0],
                     [0,0,0,0,0]],
```

Summary

In this chapter you learned how to use different display devices to show important information used in embedded design projects. In Chapter 6, we will look at audio projects using audio playback and monitoring.

Audio Projects

THIS CHAPTER IS ALL ABOUT AUDIO—both input and output. Unlike other embedded devices, the BeagleBone does not have any built-in analog audio output; however, there is digital audio out through HDMI and USB for a sound card. The BeagleBone Black drives the audio output from an I2S interface on the AM3359, which sends data to an NXP TDA1998 for conversion to HDMI. Luckily for us, the I2S interface is featured in the pin headers, as detailed in Table 6-1.

TABLE 6-1	I2S Interface Pins	
Label	**Header Pin**	**Description**
SPI1_CS0	P9_28	Bitstream
SPI1_D0	P9_29	Left/right clock
SPI1_SCLK	P9_31	Bit clock

Project 23
Internet Radio

Project 23 uses the Python language to stream Internet radio channels using the BeagleBone Black. This project does not require any additional hardware; for this example, we will output the audio through the default audio output device (HDMI, in this case).

Hardware

For this project, we not going to use any additional hardware and instead will concentrate more on the software side of things. The BeagleBone Black has default audio output on HDMI, which in theory is not very practical because this requires an HDMI device for the audio. Also, we cannot easily embedded this device into the project. Alternatively, we can use a simple USB sound card in the USB port on the BeagleBone Black.

Software

Before we get started, you need to install a media player to play streamed music on the BeagleBone Black. We are going to use mplayer, which is a movie player program that runs on a variety of formats, including Angstrom and Debian. To install mplayer, type the following in the Terminal window:

- Angstrom   `opkg install mplayer`
- Debian   `apt-get install mplayer`

If you want to run the program from the command line, you can do so using the following command:

```
mplayer —playlist http://www.bbc.co.uk/
radio/listen/live/r1_aaclca.pls
```

There are lots of free streaming radio stations on the Internet in the United States, such as www.usliveradio.com, and in the United Kingdom there is www.listenlive.eu/uk.html. Both offer a large selection of radio stations. The best way to get the URL for the streaming radio station is to right-click the link and select Copy Link Address.

The stream plays well in mplayer, but doing so is not practical in real terms because you have to stop and start the process again if you want to change the radio station to another stream. We can create a simple script using Python allows us to list the radio stations and prompt the user to select which station to listen to. Then, when the user wants to change the station, they simply select a different option in the menu. Here is the Python sketch for Project 23:

```python
#!/usr/bin/python

import subprocess
import time
import sys
import os

radio1 = 'http://www.bbc.co.uk/radio/
    listen/live/r1_aaclca.pls'
radio2 = 'http://www.bbc.co.uk/radio/
    listen/live/r2_aaclca.pls'
radio4 = 'http://www.bbc.co.uk/radio/
    listen/live/r4_aaclca.pls'
cap = 'http://media-ice.musicradio.com/
    CapitalMP3.m3u'
null = 'Nothing'

stations =
    [null,radio1,radio2,radio4,cap]

def example_1(n):
        for i in range(n):
                time.sleep(0.3)
                print '\b.',
                sys.stdout.flush()
        print ' Playing!'
        time.sleep(2)

while True:
    os.system('clear')
    print 'Please choose Radio station:'
    print '1. Radio 1'
    print '2. Radio 2'
    print '3. Radio 4'
    print '4. Capitol FM'
```

```python
    option = input('Select your radio
        station: ')

    option = int(option)

    if option > 0:

        print 'This is the number you
            selected: ' + str(option)

        print stations[(option)]

        station = stations[(option)]
        subprocess.call("mplayer
            -playlist " + station + " &",
            shell=True,
            stdout=subprocess.PIPE,
            stderr=subprocess.PIPE)
        print 'Initializing ',
            example_1(10)
        time.sleep (3)
```

In order to run the mplayer program in Python, we use a module called "subprocess." This module allows us to run the mplayer program with the given arguments as a secondary process.

The first step of the program is to list each radio station's URL in a variable so we can prompt the user to input a number. Here's an example:

```python
radio1 = 'http://www.bbc.co.uk/radio/
    listen/live/r1_aaclca.pls'
```

The variables then get put into a string, so when the user selects the number of the radio station, they are actually calling the URL from the string. You will notice that the first variable in the string is null; this is because the string starts with 0. Note that when the user selects an option from the menu, they select 1–4, but in fact, the options are 0–3. Therefore, by adding a dummy value, we can skip the first variable in the string. So, now when we select the option, we select strings 1-4:

```python
stations =
    [null,radio1,radio2,radio4,cap]
```

We now display the radio station options by using `print`. But before we do, we clear the display so that when we loop the options, the previous screen is removed from the CLI:

```
while True:
    os.system('clear')
    print 'Please choose Radio station:'
    print '1. Radio 1'
    print '2. Radio 2'
    print '3. Radio 4'
    print '4. Capitol FM'

    option = input('Select your radio
        station: ')
```

When the user inputs their selection, it gets stored in the variable `option`. We then get the URL from the string and store this in `station`:

```
station = stations[(option)]
```

We now have the selection and the URL, so we can run the mplayer program with the arguments using `subprocess`:

```
subprocess.call("mplayer -playlist " +
    station + " &", shell=True,
    stdout=subprocess.PIPE,
    stderr=subprocess.PIPE)
```

Project 24
"The Imperial March" Player

Project 24 plays a series of musical notes through a miniature speaker using the PWM pins on the BeagleBone Black. You can generate sounds from the BeagleBone Black simply by turning one of the GPIO pins on and off at the right frequency. Doing this will create a square wave, which can sound very rough. To produce a clearer sound, you need a signal that is more like a sine wave (see Figure 6-1).

To create a sine wave, we can use the analog out pins on the BeagleBone Black to output a waveform. The analog outputs from the BeagleBone Black are technically not true analog signals but PWM outputs that turn on

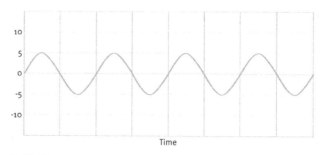

Figure 6-1 Sine wave

and off at a high frequency rate. This frequency rate is also our audio frequency. If we left the circuit untouched, the sound would be unclear and would sound just as bad as the square wave.

Hardware

Table 6-2 lists the components and equipment required for Project 24.

TABLE 6-2	Components and Equipment for Project 24	
Schematic Reference	Description	Appendix A
	BeagleBone Black	M1
	Solderless breadboard	H2
	Jumper wires	H1
H1	8 ohm speaker	H12
R2	10K POT	R4
C1	100uF capacitor	C2
C2	100nF capacitor	C1
R1	470 ohm resistor	R1
	1W IC AMP	S8

For this project we will keep the hardware to a minimum. To do this, we will use an integrated circuit 1W amplifier to amplify the signal and drive the speaker. The TDA2822M provides up to 1W of power output to the speaker in the form of an eight-pin DIP. You can see a detailed circuit diagram in Figure 6-2 and the breadboard layout in Figure 6-3. The completed Project 24 is shown in Figure 6-4.

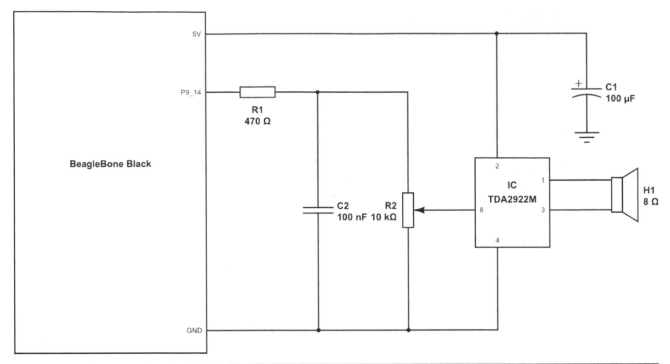

Figure 6-2 Project 24 schematic

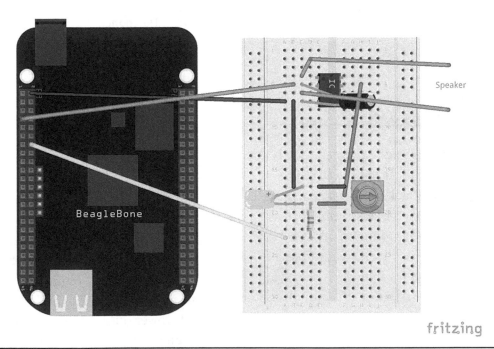

Figure 6-3 Breadboard layout for Project 24

Figure 6-4 Project 24 "Imperial March" player

Both R1 and C1 make a low-pass filter that will filter out the high-frequency PWM noise before it gets passed to the amplifier chip. The C2 capacitor is used as a decoupling capacitor that gets rid of any unwanted noise on the power rails to ground. Finally, the variable resistor R2 is a potential divider to reduce the signal from the resistor ladder; essentially, this will act as a volume control dial.

Software

To play a note using the PWM, we first must know the frequency of that note so we can set it using the PWM command. We can see on the following website that each note has a frequency in hertz:

```
www.phy.mtu.edu/~suits/notefreqs.html
```

We can set each note as variable and then call the note we want to play in the PWM command. Being that we're Evil Geniuses, the following code will play part of "The Imperial March" from *Star Wars*:

```
#!/usr/bin/python
import Adafruit_BBIO.PWM as PWM
from time import sleep


# Middle C

C = 261.63
D = 293.66
Db = 277.18
E = 329.63
Eb = 311.13
F = 349.23
G = 392
Gb = 369.99
A = 440
Ab = 415.30
B = 493.88
Bb = 466.16


#High C
HD = 587.33
HEb = 622.25
```

```
HG = 783.99

#LOW C
LG = 196.00

while True:
    PWM.start("P9_14", 50, G, 1)
    sleep(0.3)
    PWM.stop("P9_14")
    PWM.start("P9_14", 50, G, 1)
    sleep(0.3)
    PWM.stop("P9_14")
    PWM.start("P9_14", 50, G, 1)
    sleep(0.3)
    PWM.stop("P9_14")
    PWM.start("P9_14", 50, Eb, 1)
    sleep(0.3)
    PWM.stop("P9_14")
    PWM.start("P9_14", 50, Bb, 1)
    sleep(0.2)
    PWM.stop("P9_14")
    PWM.start("P9_14", 50, G, 1)
    sleep(0.3)
    PWM.stop("P9_14")
    PWM.start("P9_14", 50, Eb, 1)
    sleep(0.3)
    PWM.stop("P9_14")
    PWM.start("P9_14", 50, Bb, 1)
    sleep(0.2)
    PWM.stop("P9_14")
    PWM.start("P9_14", 50, G, 1)
    sleep(0.3)
    PWM.stop("P9_14")
    PWM.start("P9_14", 50, HD, 1)
    sleep(0.3)
    PWM.stop("P9_14")
    PWM.start("P9_14", 50, HD, 1)
    sleep(0.3)
    PWM.stop("P9_14")
    PWM.start("P9_14", 50, HD, 1)
    sleep(0.3)
    PWM.stop("P9_14")
    PWM.start("P9_14", 50, HEb, 1)
    sleep(0.3)
    PWM.stop("P9_14")

    PWM.start("P9_14", 50, Bb, 1)
    sleep(0.3)
    PWM.stop("P9_14")

    PWM.start("P9_14", 50, G, 1)
    sleep(0.3)
    PWM.stop("P9_14")
    PWM.start("P9_14", 50, Eb, 1)
    sleep(0.3)
    PWM.stop("P9_14")
    PWM.start("P9_14", 50, Bb, 1)
    sleep(0.3)
    PWM.stop("P9_14")
    PWM.start("P9_14", 50, LG, 1)
    sleep(0.3)
    PWM.stop("P9_14")
    PWM.start("P9_14", 50, LG, 1)
    sleep(0.3)
    PWM.stop("P9_14")
    PWM.start("P9_14", 50, HG, 1)
    sleep(0.3)
    PWM.stop("P9_14")
    PWM.start("P9_14", 50, HG, 1)
    sleep(0.3)
    PWM.stop("P9_14")
    PWM.start("P9_14", 50, Gb, 1)
    sleep(0.3)
    PWM.stop("P9_14")
    PWM.start("P9_14", 50, F, 1)
    sleep(0.3)
    PWM.stop("P9_14")
    PWM.start("P9_14", 50, E, 1)
    sleep(0.3)
    PWM.stop("P9_14")
    PWM.start("P9_14", 50, Eb, 1)
    sleep(0.3)
    PWM.stop("P9_14")
    PWM.start("P9_14", 50, E, 1)
    sleep(0.3)
    PWM.stop("P9_14")
    PWM.start("P9_14", 50, Ab, 1)
    sleep(0.3)
    PWM.stop("P9_14")
    PWM.start("P9_14", 50, Db, 1)
    sleep(0.3)
    PWM.stop("P9_14")
    PWM.start("P9_14", 50, C, 1)
    sleep(0.3)
    PWM.stop("P9_14")
    PWM.start("P9_14", 50, B, 1)
    sleep(0.3)
    PWM.stop("P9_14")
    PWM.start("P9_14", 50, Bb, 1)
```

```
sleep(0.3)
PWM.stop("P9_14")
PWM.start("P9_14", 50, A, 1)
sleep(0.3)
PWM.stop("P9_14")
PWM.start("P9_14", 50, Bb, 1)
sleep(0.3)
PWM.stop("P9_14")
PWM.start("P9_14", 50, Eb, 1)
sleep(0.3)
PWM.stop("P9_14")
PWM.start("P9_14", 50, Gb, 1)
sleep(0.3)
PWM.stop("P9_14")
PWM.start("P9_14", 50, Bb, 1)
sleep(0.3)
PWM.stop("P9_14")
PWM.start("P9_14", 50, G, 1)
sleep(0.3)
PWM.stop("P9_14")

PWM.cleanup()
sleep(10)
```

Project 25
Audio Level Indicator

Project 25 is used to analyze the sound levels in an environment and output the reading to a multicolored LED bar. When the noise levels are low, the green LEDs on the progress bar will light up, and as the noise levels increase, the LEDs will light up one by one to the top, where a red LED will be lit to indicate the noise levels are too high.

Hardware

Table 6-3 lists the components and equipment required for Project 25.

This project uses the Adafruit electret microphone amplifier breakout board based on the MAX9814 IC. The breakout board comes fully assembled with a 20Hz–20KHz electret microphone soldered on. Basically, an electret microphone uses a stable dielectric material

TABLE 6-3	Components and Equipment for Project 25	
Schematic Reference	Description	Appendix A
	BeagleBone Black	M1
	Jumper wires	H1
	Solderless breadboard	H2
M1	Electret mic	M8
R1	10K resistor	R5
R2	3.3K resistor	R9
C1	100nF	C1
D1	LED bar array	

that does not decay and uses electrostatic and magnets to draw an analog signal (see Figure 6-5). Because the electret module uses a 3.3V circuit, we create a simple voltage divider because we are going to read the voltage values using the BeagleBone Black's analog input, which is 1.8V maximum. Looking at the breadboard layout diagram (Figure 6-6), you can see that both the 10K and 3.3K resistors will form a voltage divider, with the center voltage around 0.8V, which is toward the middle of the BeagleBone Black's ADC range. The 100nF capacitor from the output of the electret microphone module will cause the voltage to

Figure 6-5 Adafruit electret microphone

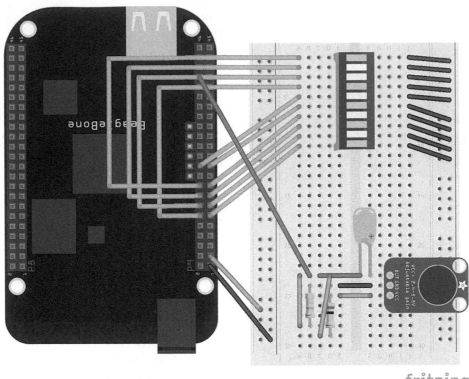

Figure 6-6 Breadboard layout diagram for Project 25

increase and decrease with the output from the microphone amplifier.

To display our output audio level, we are going to use an LED array bar, which has 10 LEDs arranged in a progress bar (see Figure 6-7). This LED bar also uses green, yellow, and red LEDs to indicate which end of the spectrum our audio levels are at. Table 6-4 shows how the LED bar array's colors are segmented.

TABLE 6-4	HDSP-4832 LED Bar Array Segment Colors
Segment	**Segment Color**
A	Red
B	Red
C	Red
D	Yellow
E	Yellow
F	Yellow
G	Yellow
H	Green
I	Green
J	Green

Back in Project 2 we discussed the anode and cathode pins in the LED. This is the same with the LED bar array, where one side of the bar is anode (+V) and the other side of the LED bar array is cathode (GND). Table 6-5 details the LED bar array's pinout.

Figure 6-7 LED bar array

NOTE Pin 1 is always indicated by a small dot on the LED bar.

TABLE 6-5	HDSP-4832 LED Bar Array Pinout		
Pin	Function	Pin	Function
1	Anode a	11	Cathode a
2	Anode b	12	Cathode b
3	Anode c	13	Cathode c
4	Anode d	14	Cathode d
5	Anode e	15	Cathode e
6	Anode f	16	Cathode f
7	Anode g	17	Cathode g
8	Anode h	18	Cathode h
9	Anode i	19	Cathode i
10	Anode j	20	Cathode j

Software

Here is the JavaScript code for Project 25:

```
var b = require('bonescript');

var ledPin0 = "P9_11";
var ledPin1 = "P9_13";
var ledPin2 = "P9_15";
var ledPin3 = "P9_17";
var ledPin4 = "P9_21";
var ledPin5 = "P9_22";
var ledPin6 = "P9_12";
var ledPin7 = "P9_14";
var ledPin8 = "P9_16";
```

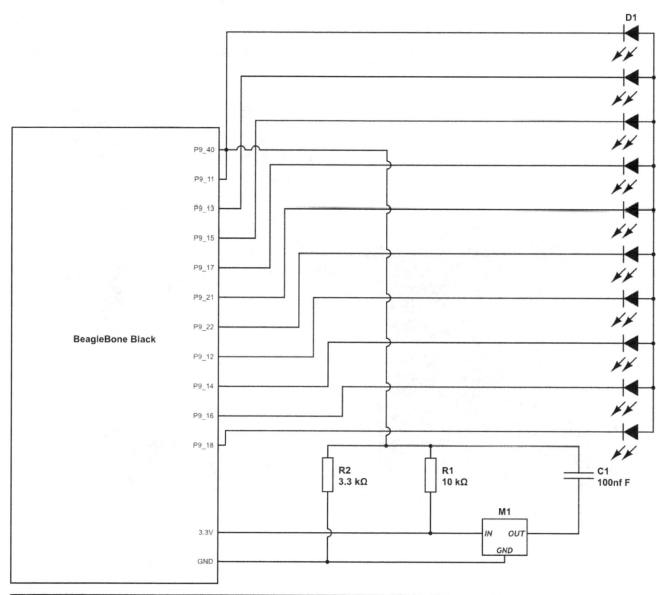

Figure 6-8 Project 25 schematic

```
var ledPin9 = "P9_18";

b.pinMode(ledPin0, b.OUTPUT);
b.pinMode(ledPin1, b.OUTPUT);
b.pinMode(ledPin2, b.OUTPUT);
b.pinMode(ledPin3, b.OUTPUT);
b.pinMode(ledPin4, b.OUTPUT);
b.pinMode(ledPin5, b.OUTPUT);
b.pinMode(ledPin6, b.OUTPUT);
b.pinMode(ledPin7, b.OUTPUT);
b.pinMode(ledPin8, b.OUTPUT);
b.pinMode(ledPin9, b.OUTPUT);

var inputPin = "P9_40";
loop()

function loop() {
    var value = b.analogRead(inputPin);
```

```
console.log(value*1.8);
setTimeout(loop, 250);
voltage = value*1.8;

if (voltage > 0.4){
    b.digitalWrite(ledPin9, b.HIGH);
}
else {
    b.digitalWrite(ledPin9, b.LOW);
}
 if (voltage > 0.5){
    b.digitalWrite(ledPin8, b.HIGH);
}
else {
    b.digitalWrite(ledPin8, b.LOW);
}
 if (voltage > 0.6){
    b.digitalWrite(ledPin7, b.HIGH);
```

Figure 6-9   Project 25 audio level indicator

The code is divided up into three main sections. The first part of the code lists all the variables for the pins, which are connected to the anode of the LED bar array. Then the pin mode is set to output so we can switch the LEDs on and off.

```
} 
else {
    b.digitalWrite(ledPin7, b.LOW);
}
 if (voltage > 0.7){
    b.digitalWrite(ledPin6, b.HIGH);
}
else {
    b.digitalWrite(ledPin6, b.LOW);
}
 if (voltage > 0.8){
    b.digitalWrite(ledPin5, b.HIGH);
}
else {
    b.digitalWrite(ledPin5, b.LOW);
}
 if (voltage > 0.9){
    b.digitalWrite(ledPin4, b.HIGH);
}
else {
    b.digitalWrite(ledPin4, b.LOW);
}
 if (voltage > 1){
    b.digitalWrite(ledPin3, b.HIGH);
}
else {
    b.digitalWrite(ledPin3, b.LOW);
}
 if (voltage > 1.1){
    b.digitalWrite(ledPin2, b.HIGH);
}
else {
    b.digitalWrite(ledPin2, b.LOW);
}
 if (voltage > 1.2){
    b.digitalWrite(ledPin1, b.HIGH);
}
else {
    b.digitalWrite(ledPin1, b.LOW);
}
 if (voltage > 1.3){
    b.digitalWrite(ledPin0, b.HIGH);
}
else {
    b.digitalWrite(ledPin0, b.LOW);
}
}
```

```
var ledPin0 = "P9_11";
var ledPin1 = "P9_13";
var ledPin2 = "P9_15";
var ledPin3 = "P9_17";
var ledPin4 = "P9_21";
var ledPin5 = "P9_22";
var ledPin6 = "P9_12";
var ledPin7 = "P9_14";
var ledPin8 = "P9_16";
var ledPin9 = "P9_18";

b.pinMode(ledPin0, b.OUTPUT);
b.pinMode(ledPin1, b.OUTPUT);
b.pinMode(ledPin2, b.OUTPUT);
b.pinMode(ledPin3, b.OUTPUT);
b.pinMode(ledPin4, b.OUTPUT);
b.pinMode(ledPin5, b.OUTPUT);
b.pinMode(ledPin6, b.OUTPUT);
b.pinMode(ledPin7, b.OUTPUT);
b.pinMode(ledPin8, b.OUTPUT);
b.pinMode(ledPin9, b.OUTPUT);
```

The second part of the code is the function, which we loop every 250 ms to get a more accurate update on the current audio levels. We read the value from the microphone and convert this into a voltage so that we know where the audio level sits with the range of 0 to 1.8V.

```
function loop() {
    var value = b.analogRead(inputPin);
    console.log(value*1.8);
    setTimeout(loop, 250);
    voltage = value*1.8;
```

For every LED in the bar array, we need to determine whether it should be on or off. The LED array has 10 LEDs, so we can set up to 10 values in the voltage range 0–1.8V. It is very unlikely that the audio levels will be 0 (low) and also unlikely that they will be 1.8V (high). If we work on the basis of a value every 100mV, then we can select a range from 0.4–1.3V, which gives us 10 values to work with. To work out whether the LED is on or off, we create an `if` statement. If the value from the microphone is greater than 0.4 (first value), we want the first LED to switch on. Otherwise, the LED will be off (if the value is less than 0.3). We can copy this calculation for all 10 LEDs.

```
if (voltage > 0.4){
        b.digitalWrite(ledPin9, b.HIGH);
    }
    else {
        b.digitalWrite(ledPin9, b.LOW);
    }
```

Summary

In this chapter we have learned about creating hardware and software audio projects using various types of hardware methods to create and monitor sound. In the next chapter, we will create some spy projects enabling us to do some evildoing and sneakiness while also protecting our evil underground lair.

CHAPTER 7

Spy Projects

This chapter isn't for the fainthearted; the more Evil Genius, the better. In the following projects, we aim to monitor and capture our target and then use interrogation techniques to determine whether the culprit is telling the truth. This chapter uses a wide range of hardware and software to create projects such as a webcam CCTV monitor, a motion detector, an automatic dog barker, and a lie detector.

Project 26
Intruder Alert
Using the Twitter API

Project 26 enables you to detect whether someone is in the room using a basic motion detector sensor. You'll use the Python Twitter API to display a tweet once you have detected an intruder.

A motion detection sensor uses a passive infrared (PIR) filter, which is an electronic sensor that measures infrared from an object or person. Because humans radiate heat, this sensor will be perfect for detecting the presence of someone in a room. The sensor detects a change in radiation given off by other objects, so the motion is triggered only when the change is higher than the normal average value.

Hardware

Table 7-1 lists the components and equipment required for Project 26.

TABLE 7-1	Components and Equipment for Project 26
Description	**Appendix A**
BeagleBone Black	M1
Solderless breadboard	H2
Jumper wires	H1
PIR sensor	S9

This project uses a simple PIR sensor that detects motion at a wide angle of 102 degrees. This gives us a better chance of detecting our target, up to a range of 12m. The sensor has three pins: +V (which ranges from 3–6VDC), the Ground pin, and an output pin (which outputs a voltage when the sensor has detected motion). The output pin is connected to one of the GPIO pins on the BeagleBone Black, so we can use this pin as an input and detect when the value is high or a voltage is present. The breadboard layout diagram for Project 26 is shown in Figure 7-1 and the completed Project 26 in Figure 7-2.

Software

Before we get started, you need to make sure you have the Twitter API program installed on your BeagleBone Black.

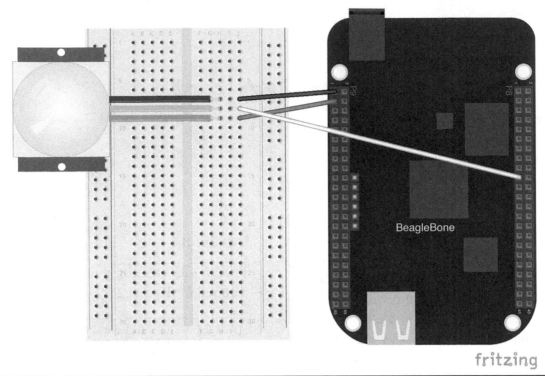

fritzing

Figure 7-1   Breadboard layout for Project 26

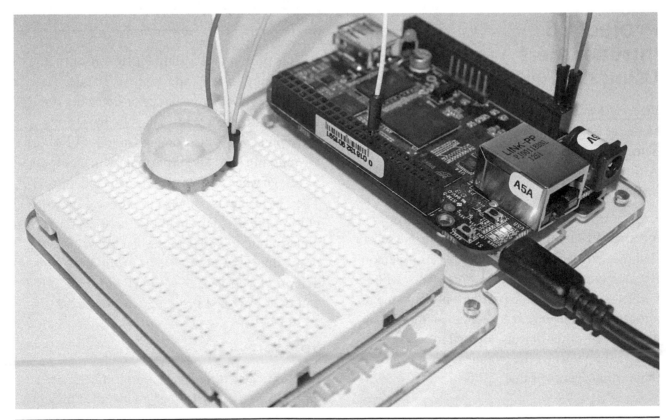

Figure 7-2   Project 26 intruder alert using the Twitter API

Debian:

- `sudo apt-get install python-setuptools`
- `sudo easy_install pip`
- `sudo pip install tweepy`

Angstrom:

- `opkg install python-setuptools`
- `easy_install tweepy`

Now that you have the Twitter API installed, you need to create a Twitter account so you can post a tweet when you detect motion. To do this, follow these steps:

1. Go to https://twitter.com/ and sign up with a new account.

2. Go to https://dev.twitter.com/ and sign in with the account you have just created.

3. In the top-right corner of the screen, click your account and go to My Applications.

4. On the next screen, click the Create New App button.

5. Fill out all the details on the screen (see Figure 7-3), including a unique name of your application, a short description of what the application is for and what it does, and the website URL where the application information can be found (if the app is

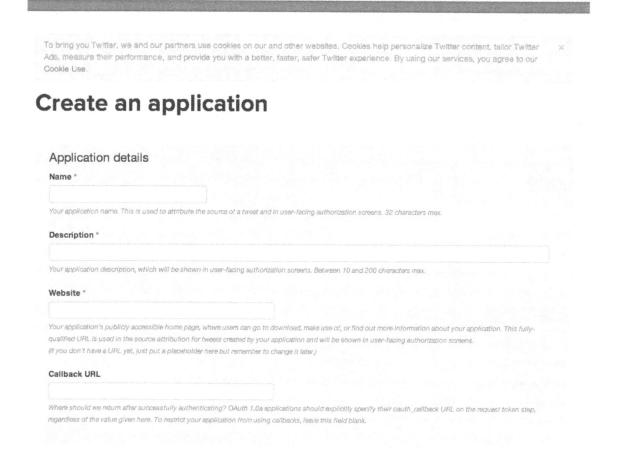

Figure 7-3 Creating an application

being used in a web page, enter this web page's URL). We are not going to use a callback function, so you can leave the Callback URL field blank.

6. Accept the terms and conditions and then click the Create Your Twitter Application button.

7. Once the app has been created, it gives you an API key, which you can use to access your Twitter account (it's almost a unique ID for your account). If you click the API Keys tab, all your API information, such as consumer key and consumer secret, is displayed as shown in Figure 7-4.

8. In order to make authorized calls to Twitter's API, the application must obtain

an OAuth access token. To do this, in the API Keys section, look for the option at the bottom of the screen to create an access token as shown in Figure 7-5.

9. Once this token is created, you should see on the API Keys page all the token information (Figure 7-6). Make sure you keep a note of all these API keys because you'll need this information in the program.

The code for this project uses a number of libraries such as Adafruit GPIO library for referencing the BeagleBone Black's pins using Python and also Tweepy, which is a Twitter Python library for sending receiving tweets. In order for use to use the Twitter API, we must insert the API keys into the variables. In the following Python code, we use

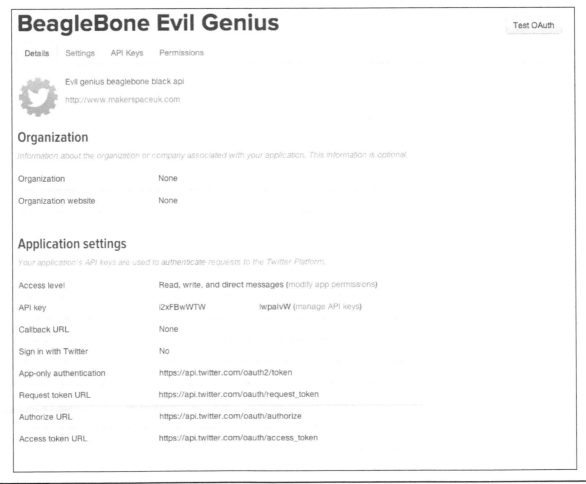

Figure 7-4 Application details

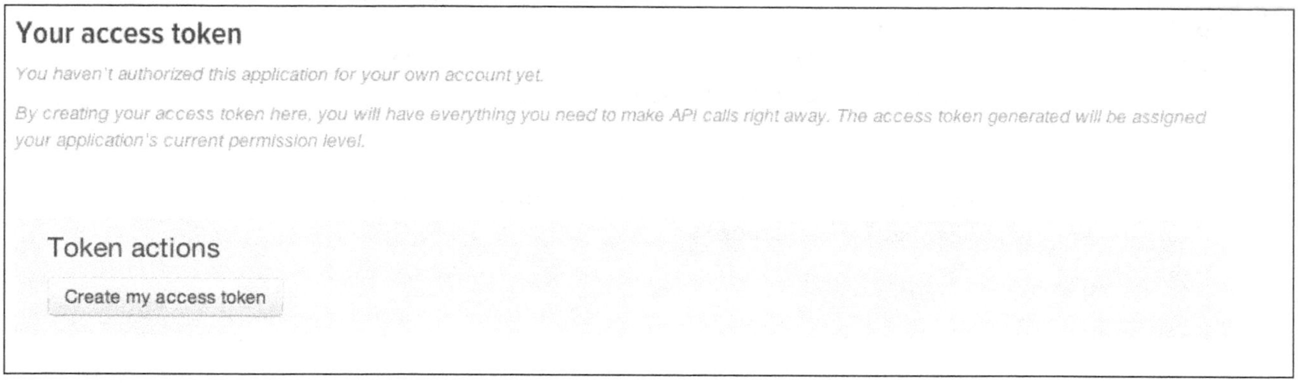

Your access token

You haven't authorized this application for your own account yet.

By creating your access token here, you will have everything you need to make API calls right away. The access token generated will be assigned your application's current permission level.

Token actions

Create my access token

Figure 7-5 Requesting an access token

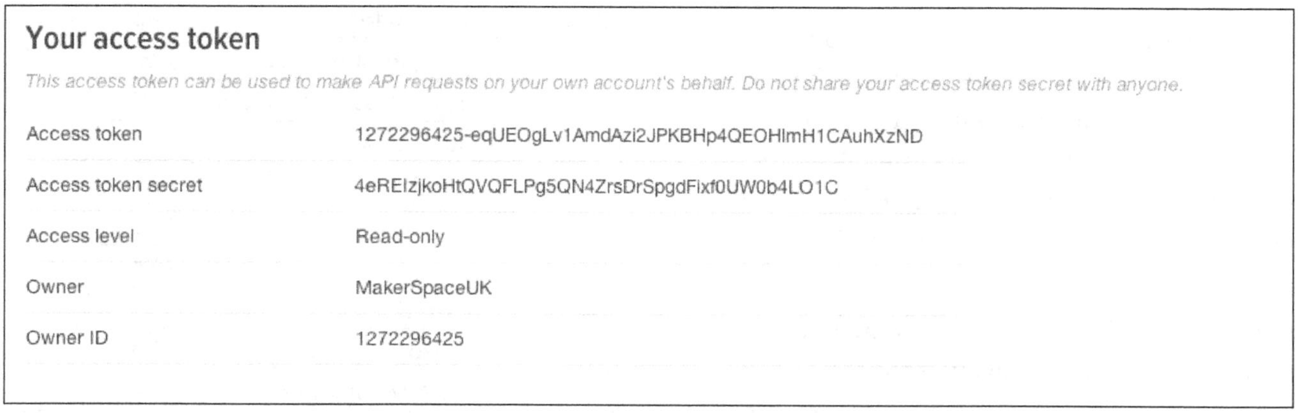

Your access token

This access token can be used to make API requests on your own account's behalf. Do not share your access token secret with anyone.

Access token	1272296425-eqUEOgLv1AmdAzi2JPKBHp4QEOHlmH1CAuhXzND
Access token secret	4eREIzjkoHtQVQFLPg5QN4ZrsDrSpgdFixf0UW0b4LO1C
Access level	Read-only
Owner	MakerSpaceUK
Owner ID	1272296425

Figure 7-6 Access token keys

"GPIO.wait_for_edge" function, which allows us to detect a signal from the PIR sensor. This method is preferred over using a simple loop function. Here is the Python code for Project 26:

```python
import Adafruit_BBIO.GPIO as GPIO
from time import sleep
import os
import sys
import tweepy

CONSUMER_KEY = 'xxxxxxxxxxxxxxxxxx'
CONSUMER_SECRET = 'xxxxxxxxxxxxxxxxxxxxxxxxxx'
ACCESS_KEY = 'xxxxxxxxxxxxxxxxxxxxxxxxxxxxxxxx'
ACCESS_SECRET = 'xxxxxxxxxxxxxxxxxxxxxxxxxxxxx'

GPIO.setup("P8_19", GPIO.IN)

while True:
    GPIO.wait_for_edge("P8_19", GPIO.RISING)
```

```
print ("Motion detected")
auth = tweepy.OAuthHandler(CONSUMER_KEY, CONSUMER_SECRET)
auth.set_access_token(ACCESS_KEY, ACCESS_SECRET)
api = tweepy.API(auth)

api.update_status('ALERT!! Motion detected in Zone 1 -
Message sent from #BeagleBoneBlack')
sleep(60)
```

Project 27
Lie Detector

So you have detected and captured your intruder—they were unable to evade the Evil Genius! Now you need to interrogate your suspect. Are they telling the truth? Well, for Project 27, we will create a simple lie detector using some simple circuitry and mathematics.

We use the intruder's hand to tell whether or not they are telling a lie. When someone gets nervous, they usually have sweaty hands and their skin's resistance begins to decrease. We can create a circuit to measure this resistance using the BeagleBone Black's analog input. We will also use a simple RGB LED and a buzzer to indicate when the intruder is telling the truth.

Hardware

Table 7-2 lists the components and equipment required for Project 27.

We use an RGB LED in this project to display whether the subject is telling a lie. When the LED turns red, this indicates a lie. Green is used to indicate the truth, and blue indicates that the detector needs slight adjusting using the variable resistor.

The circuit works by using the subject as a resistor, measuring their skin resistance and

TABLE 7-2	Components and Equipment for Project 27	
Schematic Reference	**Description**	**Appendix A**
	BeagleBone Black	M1
	Solderless breadboard	H2
	Jumper wires	H1
R1, 2, 3	220R resistors	R1
R5	10K resistor	R5
R4	10K POT	R4
D1	RGB LED	S3
H1	Piezoelectric buzzer	H13
	Paper clips/thumbtacks	

creating a voltage divider with a fixed resistor on one end. The lower the intruder's resistance, the more the analog input will be pulled to 1.8V; the higher the resistance, the further toward ground it will be pulled.

The variable resistor is used to compare and adjust the point of resistance to ±10 percent of the two contact points. Therefore, when the resistance of the contact point falls below the variable resistor value, the detector needs adjusting to within ±10 percent of the value. For a more detailed explanation, see Figure 7-7; the breadboard layout can be seen in Figure 7-8. The completed Project 27 is shown in Figure 7-9.

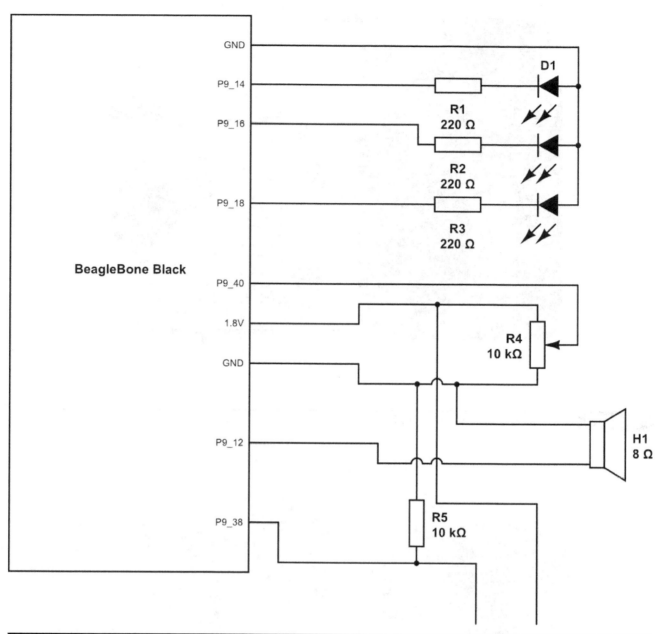

Figure 7-7 Project 27 schematic diagram

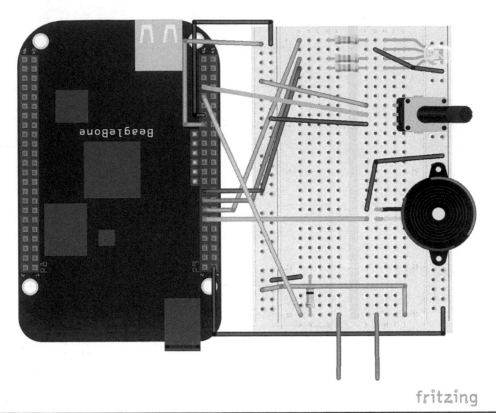

Figure 7-8   Breadboard layout for Project 27

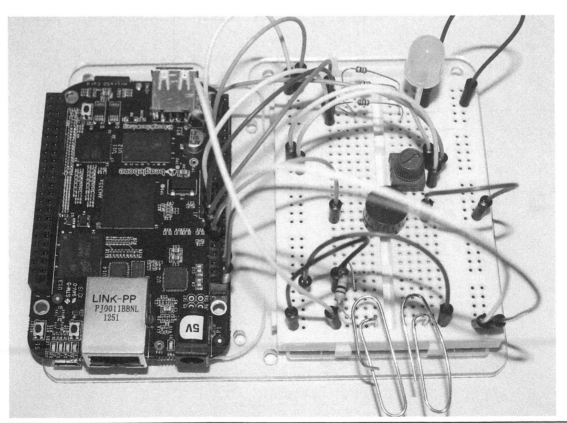

Figure 7-9   Project 27 lie detector

Software

The script for Project 27 compares the voltage for both A1 and A0. If they are about the same, the RGB LED will be green. If the voltage from A0 is 10 percent higher than A1, the RGB LED will change to red and the buzzer will sound to indicate the suspect is telling a lie. If A1 is less than A0, the RGB LED will turn blue, indicating that the variable adjuster needs changing until the LED turns green.

```python
import Adafruit_BBIO.GPIO as GPIO
import Adafruit_BBIO.ADC as ADC
from time import sleep
GPIO.setup("P9_12", GPIO.OUT) #buzzer

GPIO.setup("P9_14", GPIO.OUT)
GPIO.setup("P9_16", GPIO.OUT)
GPIO.setup("P9_18", GPIO.OUT)

while True:
    def beep():
        GPIO.output("P9_12", GPIO.HIGH)
        sleep(1)
        GPIO.output("P9_12", GPIO.LOW)
        sleep(1)

    ADC.setup()
    value = ADC.read("P9_40")
    value2 = ADC.read("P9_38")
    voltagePOT = value * 1.8 #1.8V
    voltagePIN = value2 * 1.8
    print (voltagePOT)
    print (voltagePIN)
    sleep(1)

    if (voltagePIN > voltagePOT * 1.1):
        GPIO.output("P9_14", GPIO.HIGH)
        GPIO.output("P9_16", GPIO.LOW)
        GPIO.output("P9_18", GPIO.LOW)
        beep()

    elif (voltagePIN < voltagePOT *
        1.1):
        GPIO.output("P9_14", GPIO.LOW)
        GPIO.output("P9_16", GPIO.HIGH)
        GPIO.output("P9_18", GPIO.LOW)
```

```python
    else:
        GPIO.output("P9_14", GPIO.LOW)
        GPIO.output("P9_16", GPIO.LOW)
        GPIO.output("P9_18", GPIO.HIGH)

    sleep(5)
```

Project 28
Keypad Door Latch

Project 28 is designed to keep your doors locked by using a standard 3×4 matrix keypad as a passcode system and a solenoid to act as a door latch for your door. When the correct code has been entered into the keypad, the door will either lock or unlock.

A DC power supply is required to be able to supply the solenoid with enough current to activate the latch. Be sure to check the datasheet for the specification of your power device because some devices vary. Normally, a 1A power supply unit (PSU) will be fine.

Hardware

Table 7-3 lists the components and equipment required for Project 28.

TABLE 7-3	Components and Equipment for Project 28
Description	**Appendix A**
BeagleBone Black	M1
Solderless breadboard	H2
Jumper wires	H1
4×3 matrix keypad	H14
Solenoid	H15
TIP120 transistor	S10
1N4004 diode	S11
DC PSU 12V 2A	H16

The solenoid in this circuit has an inductive load, so it is liable to generate backward electromotive force (EMF). This is why we include a diode in the circuit because this protects the circuit against such an effect and prevents damage to the board and other components.

The solenoid is controlled using the TIP120 transistor. The transistor acts as a switch using three pins: the base, collector, and emitter. The base pin is connected to the BeagleBone Black, and when this pin is high, it switches the current from the collector to the emitter, thus creating the circuit.

The keypad is a standard 3×4 keypad with seven output pins that represent either a column or a row. Every keypad may have a slightly different pinout, so always refer to the datasheet for exactly how to wire it up. We use the keypad by scanning each column and then each row until we detect a value; then we know exactly which number has been pressed. The breadboard

layout diagram can be seen in Figure 7-10, and the completed project is shown in Figure 7-11.

Software

The code for this project scans each row and column and detects whether any of the pins is high. To do this, we alternate between scanning the rows and columns, setting each to HIGH or LOW.

We ask the user to enter four digits on the keypad and then we check whether these digits match our passcode, which is stored as a variable. If the passcode is correct, we grant access and turn the latch either on or off, depending on its current state. If the passcode is incorrect, we prompt the user to enter the correct passcode again.

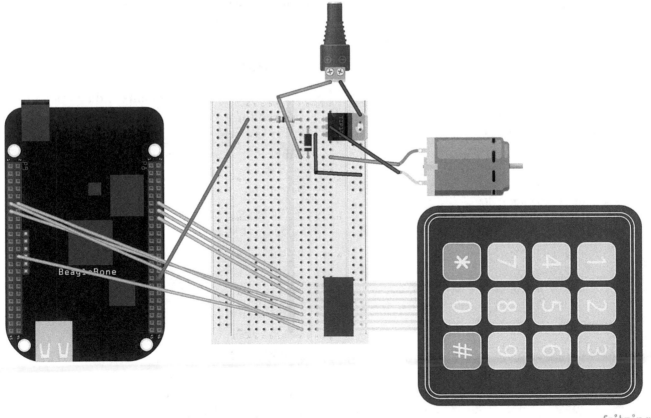

Figure 7-10   Breadboard layout for Project 28

Figure 7-11   Project 28 keypad door latch

```
import Adafruit_BBIO.GPIO as GPIO
from time import sleep

class keypad():
    def __init__(self, columnCount = 3):
        pin1 = "P8_14"
        pin2 = "P8_16"
        pin3 = "P8_11"
        pin4 = "P9_13"
        pin5 = "P9_12"
        pin6 = "P9_26"
        pin7 = "P9_11"
        pin8 = "P9_24"

        if columnCount is 3:
            self.KEYPAD = [
                [1,2,3],
                [4,5,6],
                [7,8,9],
                ["*",0,"#"]
            ]
            self.ROW          = [pin7, pin6, pin5, pin4]
            self.COLUMN       = [pin3, pin2, pin1]
        elif columnCount is 4:
```

```
                self.KEYPAD = [
                    [1,2,3,"A"],
                    [4,5,6,"B"],
                    [7,8,9,"C"],
                    ["*",0,"#","D"]
                ]
                self.ROW        = [pin8, pin7, pin6, pin5]
                self.COLUMN     = [pin4, pin3, pin2, pin1]
            else:
                return
    def getKey(self):
        for j in range(len(self.COLUMN)):
            GPIO.setup(self.COLUMN[j], GPIO.OUT)
            GPIO.output(self.COLUMN[j], GPIO.LOW)
        for i in range(len(self.ROW)):
            GPIO.setup(self.ROW[i], GPIO.IN, GPIO.PUD_DOWN)

        rowVal = -1
        for i in range(len(self.ROW)):
            tmpRead = GPIO.input(self.ROW[i])
            if tmpRead == 0:
                rowVal = i
        if rowVal <0 or rowVal >3:
            self.exit()
            return
        for j in range(len(self.COLUMN)):
                GPIO.setup(self.COLUMN[j], GPIO.IN, GPIO.PUD_DOWN)
        GPIO.setup(self.ROW[rowVal], GPIO.OUT)
        GPIO.output(self.ROW[rowVal], GPIO.HIGH)

        colVal = -1
        for j in range(len(self.COLUMN)):
            tmpRead = GPIO.input(self.COLUMN[j])
            if tmpRead == 1:
                colVal=j


        if colVal <0 or colVal >2:
            self.exit()
            return


        self.exit()
        return self.KEYPAD[rowVal][colVal]


    def exit(self):
        for i in range(len(self.ROW)):
                GPIO.setup(self.ROW[i], GPIO.IN, pull_up_down=GPIO.PUD_UP)
```

```python
        for j in range(len(self.COLUMN)):
            GPIO.setup(self.COLUMN[j], GPIO.IN, pull_up_down=GPIO.PUD_UP)


while True:

    kp = keypad(columnCount = 3)

    def digit():
        r = None
        while r == None:
            r = kp.getKey()
        return r
    print "Please enter a 4 digit code: "

    d1 = digit()
    print d1
    sleep(1)


    d2 = digit()
    print d2
    sleep(1)


    d3 = digit()
    print d3
    sleep(1)


    d4 = digit()
    print d4
    sleep(1)

if ("%s%s%s%s"%(d1,d2,d3,d4) == "1234"):
        print "You Entered %s%s%s%s "%(d1,d2,d3,d4)
        print "Access granted unlocking......"
        GPIO.output("P8_32", GPIO.HIGH)
        sleep(5)
        GPIO.output("P8_32", GPIO.LOW)
        sleep(1)
        GPIO.cleanup()
    else:
        print "Access denied"
        print "You Entered %s%s%s%s "%(d1,d2,d3,d4)
```

Project 29
Webcam Security Doorbell

Project 29 uses a standard Logitech webcam to take pictures when someone is at your front door. This is a great project for when you are not home and are wondering who stopped by. We use a simple switch to act as the doorbell (alternatively, you could use a vibration detector to detect when someone knocks at the door). Once the switch has been pushed, the webcam will start a stream and take a picture; we will then push this to Twitter as a post using the Tweepy Twitter API.

Hardware

Table 7-4 lists the components and equipment required for Project 29.

This project uses a C270 Logitech webcam as the security cam feed so we can capture an

TABLE 7-4	Components and Hardware for Project 29
Description	**Appendix A**
BeagleBone Black	M1
Solderless breadboard	H2
Jumper wires	H1
10K resistor	R5
Tactile switch	H3
C270 Logitech webcam	H16

image. This webcam can be powered directly from the BeagleBone Black, but just to make sure, always power the BeagleBone through a PSU. Alternatively, you can power the webcam from a powered USB hub.

 NOTE **Always connect USB devices before booting the BeagleBone Black.**

The breadboard layout diagram is shown in Figure 7-12. The resistor in the circuit is used as a pull-down resistor to prevent a de-bouncing

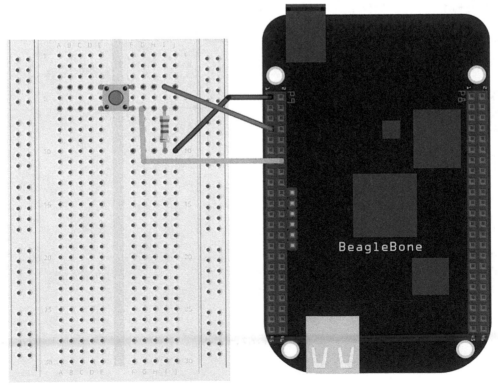

Figure 7-12 Breadboard layout for Project 29

Figure 7-13 Project 29 doorbell security

effect. The completed project can be seen in Figure 7-13.

Software

Before getting started, you must ensure that Tweepy is installed. Refer to Project 26 to see detailed instructions on how to install this API. Make sure you put all your Twitter API details for your account where all the x characters are.

In order to use the webcam, we need a streaming program that handles the video output for us. For this project, we are going to use mjpg-streamer, which takes JPG images from Linux UVC-compatible webcams and then streams them as an M-JPEG through an HTTP web browser. This allows us to view the stream over the Internet or local network. To

install mjpg-streamer, type the following into the command line:

```
Git clone git://github.com/makerspaceuk/
    mjpg-streamer.git

cd mjpg-streamer

make

sudo make install
```

Now that you have installed the mjpg-streamer, let's run a quick test just to make sure everything is working as it should be. In the same directory as /mjpg-streamer, type the following:

```
Sudo ./mjpg_streamer —I "./input_uvc.so"
—o "./output_http.so —p 8090 —w ./www"
```

Depending on the webcam, the program will output the messages in the Terminal window for configuration. Next, go to your browser and in the URL bar type **http://192.168.7.2:8090**. You should see the mjpg-streamer web page with a view from your webcam. This web page also shows you some other features of embedding the stream into your website (see Figure 7-14).

The code for Project 29 is similar to that of Project 25, with the exception of using a USB webcam to embed an image into the tweet. Our program downloads the image from the mjpg streamer program using the USB webcam and places it in a temporary folder where our Twitter API can access it. Here is the Python code for Project 29:

Figure 7-14 The mjpg-streamer web page

```
import tweepy
from subprocess import call
from datetime import datetime
from time import sleep
import subprocess
import Adafruit_BBIO.GPIO as GPIO

GPIO.setup("P9_14", GPIO.IN)

while True:
    i = datetime.now()
    now = i.strftime('%Y%m%d-%H%M%S')
    GPIO.wait_for_edge("P9_14", GPIO.RISING)
    kill = 'killall mjpg_streamer'
    cmd = 'wget -O /root/frontdoor.jpg http://192.168.7.2:8090/?action=snapshot'
    webcam = '/root/mjpg-streamer/mjpg_streamer -i "/root/mjpg-
    streamer/input_uvc.so" -o "/root/mjpg-streamer/output_http.so -p 8090
    -w ./www" &'

call([webcam], shell=True)
    sleep(5)
    call ([cmd], shell=True)
```

```
consumer_key = 'xxxxxxxxxxxxx'
consumer_secret = 'xxxxxxxxxxxxxxx'
access_token = xxxxxxxxxxxxxxxxx'
access_token_secret = 'xxxxxxxxxxxxxxxxxx'

auth = tweepy.OAuthHandler(consumer_key, consumer_secret)
auth.set_access_token(access_token, access_token_secret)

authentication
api = tweepy.API(auth)


photo_path = '/root/123.jpg'
status = 'BeagleBone Black door bell: ' + i.strftime('%Y/%m/%d %H:%M:%S
#beaglebone')
api.update_with_media(photo_path, status=status)
call ([kill], shell=True)
```

Project 30
Automatic Dog Barker

Our final project, Project 30, is designed to keep those unwanted guests away from your property by using the HC-SR04 ultrasonic sensor and the USB audio card. The sensor detects when someone is walking up to your property and initializes an audio file of some aggressive dogs barking. This will be enough for anyone to think twice before approaching your property.

Hardware

Table 7-5 lists the components and equipment required for Project 30.

TABLE 7-5	Components and Equipment for Project 30
Description	Appendix A
BeagleBone Black	M1
Solderless breadboard	H2
Jumper wires	H1
HC-SR04 sensor	M9
USB audio card	H17

This project uses a standard HC-SR04 ultrasonic sensor to detect the distance of objects. The sensor can detect objects between 2cm and 400cm away and can measure with an accuracy of 3mm. The sensor has four pins to connect to the BeagleBone Black: 5V supply, 0V ground, trigger pulse input, and echo pulse output. The echo pulse is used to send out a signal, and the trigger is used to detect the signal coming back so that we can then work out the timing between pulses. This will give us the distance of the object.

For the audio output, we are going to use a USB audio card, which we can plug directly into the BeagleBone Black. When you first power up your BeagleBone Black, make sure the USB audio card is plugged in; otherwise, the operating system will not detect it. The breadboard layout diagram is shown in Figure 7-15, and the completed project can be seen in Figure 7-16.

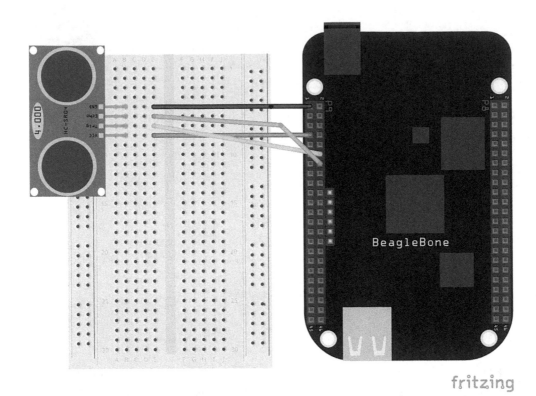

Figure 7-15 Breadboard layout for Project 30

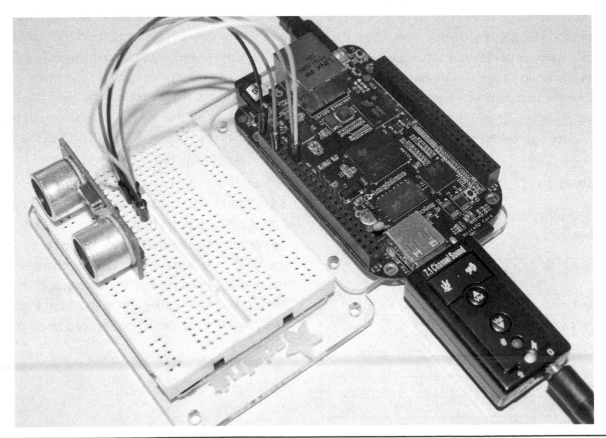

Figure 7-16 Project 30 automatic dog barker

Software

This program uses mplayer to play back the audio MP3 file of a dog barking. To install mplayer on the BeagleBone Black, type the following command:

- Angstrom `opkg install mplayer`
- Debian `apt-get install mplayer`

To play an MP3 file using mplayer, simply type `mplayer [filename].mp3`, where `[filename]` is the name of your audio file. Mplayer will start to play the audio through the default audio device, which in this case automatically comes through the HDMI. To change the audio card, you use certain parameters with mplayer. On the command line, type `alsa —L`, which gives you a list of possible audio devices to use:

```
default:CARD=Black
    TI BeagleBone Black,
    Default Audio Device
sysdefault:CARD=Black
    TI BeagleBone Black,
    Default Audio Device
default:CARD=Device
    USB PnP Sound Device, USB Audio
    Default Audio Device
sysdefault:CARD=Device
    USB PnP Sound Device, USB Audio
    Default Audio Device
```

To test the audio card, type the following:

```
speaker-test —D default:Device
```

If you hear some white noise, then great! You know it works. Now let's play an MP3 audio file using mplayer with the given parameters for the audio card. Type the following to play the file:

```
mplayer —ao alsa:device=default=Device
test.mp3
```

The audio file should now be playing. You may notice that the sound volume is very low, because this is the default setting. Change the volume type to alsamixer and then press F6 to select the USB audio card.

```python
import Adafruit_BBIO.GPIO as GPIO
import time
from subprocess import call

trigger_pin = 'P9_12'
echo_pin = 'P9_14'


GPIO.setup("P9_12", GPIO.OUT)
GPIO.setup("P9_14", GPIO.IN)

def send_trigger_pulse():
    GPIO.output("P9_12", True)
    time.sleep(0.0001)
    GPIO.output("P9_12", False)

def wait_for_echo(value, timeout):
    count = timeout
    while GPIO.input("P9_14") != value
      and count > 0:
        count = count - 1

def get_distance():
    send_trigger_pulse()
    wait_for_echo(True, 10000)
    start = time.time()
    wait_for_echo(False, 10000)
    finish = time.time()
    pulse_len = finish - start
    distance_cm = pulse_len / 0.000058
    #distance_in = distance_cm / 2.5
    #return (distance_cm, distance_in)
    return (distance_cm)

while True:
    if (get_distance() < 10):
        bark = 'mplayer -ao
alsa:device=default=Device
  /root/dog.mp3'
        call([bark], shell=True)
        time.sleep(5)
```

Summary

This was the final chapter of Evil Genius projects. I hope you have fully enjoyed creating them and have learned a lot about the great features of the BeagleBone Black. Hopefully you can take what you have learned and start creating some of your own Evil Genius projects using your appetite for experimentation and design. The next chapter gives a brief overview on how to use some basic electronic equipment and how to read datasheets, which are vital ingredients for an Evil Genius.

CHAPTER 8

Tools and Tips

THIS CHAPTER PROVIDES SOME USEFUL TIPS for creating your own projects and how to best utilize your resources. Starting your own project can be a bit daunting at first, and sometimes can seem frustrating and complicated. These tools and tips should help you on your way—and in no time, you will become an efficient Evil Genius.

Datasheets

Datasheets are manuals from the manufacturer for its electronics components. They explain exactly how a particular component should work and how it should be used. Usually, these documents are written by electrical engineers and scientists. This is why they can sometimes cause confusion among average users, especially newcomers. However difficult they may be to read, they are also the bread and butter of manuals, and sometimes they can be a key resource of information that provides all the details for a design circuit.

The actual manuscript of a datasheet will vary depending on what type of electronic part it is and who manufactured it. The first page of a typical datasheet for an integrated circuit usually provides a brief description of exactly what the product is and what its features are (see Figure 8-1). Usually, you can tell within the first few pages whether this is the right part for you. A datasheet should give you a good indication whether or not a part will work in your project.

A datasheet for an integrated circuit (IC) usually contains a pinout of the part's pins that displays their functions and where they are physically located on the chip (see Figure 8-2). There is always a indicator on the diagram as to where the first pin is on the circuit; this is very important because wiring the circuit the wrong way can cause damage to both the chip and the circuitry. The pins are all labeled using short acronyms, which are explained later in the datasheet. For example, Vcc is the supply voltage (usually a 5V or 3.3V), GND is ground, and EN is the pin that enables a certain function of the IC (usually, it prepares the circuit to receive data). If you are connecting an I2C device, you will see SCL and SDA, which indicate the clock and the data pins, respectively.

Looking further into the datasheet, you will come across detailed information of the specification, represented in tables (see Figure 8-3). These values determine the minimum and maximum ratings the component can withstand before being damaged. It is usually advisable that you keep within these restrictions; otherwise, you risk shortening the life span of the component or even damaging it and the circuit it is in.

- Featuring Unitrode L293 and L293D
  Products Now From Texas Instruments
- Wide Supply-Voltage Range: 4.5 V to 36 V
- Separate Input-Logic Supply
- Internal ESD Protection
- Thermal Shutdown
- High-Noise-Immunity Inputs
- Functionally Similar to SGS L293 and
  SGS L293D
- Output Current 1 A Per Channel
  (600 mA for L293D)
- Peak Output Current 2 A Per Channel
  (1.2 A for L293D)
- Output Clamp Diodes for Inductive
  Transient Suppression (L293D)

description/ordering information

The L293 and L293D are quadruple high-current half-H drivers. The L293 is designed to provide bidirectional drive currents of up to 1 A at voltages from 4.5 V to 36 V. The L293D is designed to provide bidirectional drive currents of up to 600-mA at voltages from 4.5 V to 36 V. Both devices are designed to drive inductive loads such as relays, solenoids, dc and bipolar stepping motors, as well as other high-current/high-voltage loads in positive-supply applications.

All inputs are TTL compatible. Each output is a complete totem-pole drive circuit, with a Darlington transistor sink and a pseudo-Darlington source. Drivers are enabled in pairs, with drivers 1 and 2 enabled by 1,2EN and drivers 3 and 4 enabled by 3,4EN. When an enable input is high, the associated drivers are enabled, and their outputs are active and in phase with their inputs. When the enable input is low, those drivers are disabled, and their outputs are off and in the high-impedance state. With the proper data inputs, each pair of drivers forms a full-H (or bridge) reversible drive suitable for solenoid or motor applications.

L293 . . . N OR NE PACKAGE
L293D . . . NE PACKAGE
(TOP VIEW)

```
1,2EN  [ 1      16 ]  V_CC1
  1A   [ 2      15 ]  4A
  1Y   [ 3      14 ]  4Y
HEAT SINK AND { [ 4      13 ] } HEAT SINK AND
  GROUND    { [ 5      12 ] }   GROUND
  2Y   [ 6      11 ]  3Y
  2A   [ 7      10 ]  3A
 V_CC2 [ 8       9 ]  3,4EN
```

L293 . . . DWP PACKAGE
(TOP VIEW)

```
1,2EN  [ 1      28 ]  V_CC1
  1A   [ 2      27 ]  4A
  1Y   [ 3      26 ]  4Y
  NC   [ 4      25 ]  NC
  NC   [ 5      24 ]  NC
  NC   [ 6      23 ]  NC
HEAT SINK AND { [ 7      22 ] } HEAT SINK AND
  GROUND      { [ 8      21 ] }   GROUND
              { [ 9      20 ] }
  NC   [ 10     19 ]  NC
  NC   [ 11     18 ]  NC
  2Y   [ 12     17 ]  3Y
  2A   [ 13     16 ]  3A
 V_CC2 [ 14     15 ]  3,4EN
```

ORDERING INFORMATION

T_A	PACKAGE†		ORDERABLE PART NUMBER	TOP-SIDE MARKING
0°C to 70°C	HSOP (DWP)	Tube of 20	L293DWP	L293DWP
	PDIP (N)	Tube of 25	L293N	L293N
	PDIP (NE)	Tube of 25	L293NE	L293NE
		Tube of 25	L293DNE	L293DNE

† Package drawings, standard packing quantities, thermal data, symbolization, and PCB design guidelines are available at www.ti.com/sc/package.

Figure 8-1 First page summary of L293DNE datasheet

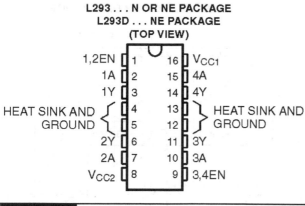

L293 . . . N OR NE PACKAGE
L293D . . . NE PACKAGE
(TOP VIEW)

Figure 8-2 L293DNE pinout

You will also see sections with the recommended operating conditions (see Figure 8-4). These may include voltage and current ranges for timings, temperature, and other useful information. These conditions are always good for setting your specification to achieve optimal performance and precision.

Some datasheets will have one or more graphs showing the component's performance versus the criteria (see Figure 8-5). The graphs displayed for the L293DNE circuit show the

electrical characteristics, V_{CC1} = 5 V, V_{CC2} = 24 V, T_A = 25°C

PARAMETER		TEST CONDITIONS		MIN	TYP	MAX	UNIT
V_{OH}	High-level output voltage	L293: I_{OH} = −1 A L293D: I_{OH} = −0.6 A		V_{CC2} − 1.8	V_{CC2} − 1.4		V
V_{OL}	Low-level output voltage	L293: I_{OL} = 1 A L293D: I_{OL} = 0.6 A			1.2	1.8	V
V_{OKH}	High-level output clamp voltage	L293D: I_{OK} = −0.6 A			V_{CC2} + 1.3		V
V_{OKL}	Low-level output clamp voltage	L293D: I_{OK} = 0.6 A			1.3		V
I_{IH}	High-level input current	A	V_I = 7 V		0.2	100	μA
		EN			0.2	10	
I_{IL}	Low-level input current	A	V_I = 0		−3	−10	μA
		EN			−2	−100	
I_{CC1}	Logic supply current	I_O = 0	All outputs at high level		13	22	mA
			All outputs at low level		35	60	
			All outputs at high impedance		8	24	
I_{CC2}	Output supply current	I_O = 0	All outputs at high level		14	24	mA
			All outputs at low level		2	6	
			All outputs at high impedance		2	4	

Figure 8-3 Specification table

L293, L293D
QUADRUPLE HALF-H DRIVERS

SLRS008C – SEPTEMBER 1986 – REVISED NOVEMBER 2004

recommended operating conditions

			MIN	MAX	UNIT
	Supply voltage	V_{CC1}	4.5	7	V
		V_{CC2}	V_{CC1}	36	
V_{IH}	High-level input voltage	$V_{CC1} \leq 7$ V	2.3	V_{CC1}	V
		$V_{CC1} \geq 7$ V	2.3	7	V
V_{IL}	Low-level output voltage		−0.3†	1.5	V
T_A	Operating free-air temperature		0	70	°C

† The algebraic convention, in which the least positive (most negative) designated minimum, is used in this data sheet for logic voltage levels.

Figure 8-4 Recommended operating conditions

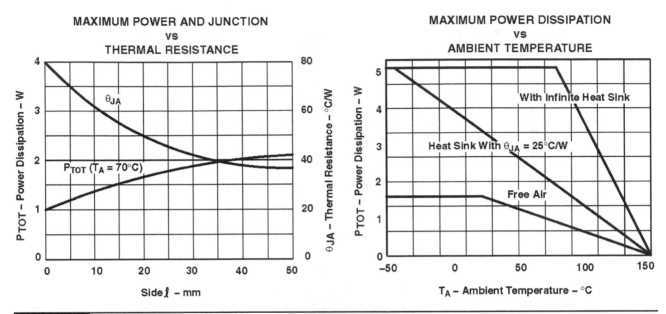

Figure 8-5   Application information

power/heat dissipation to give us an idea of how hot the chip can get when it uses a certain amount of power.

Truth tables show how changing the inputs to a part will affect the output (see Figure 8-6). Each row has all the components' inputs set to a specific state and then the resulting effect. H means the input logic is "high," and L means the logic is low or GND. If there is an X in the logic state, it doesn't matter whether the state is high or low; the output will be the same either way.

Timing diagrams show clearly how data should be transmitted and received from the component as well as at which speed this should be done (see Figure 8-7). These are typically represented as horizontal lines with various

inputs and outputs; the horizontal line shows the logic over a period of time. If the line rises higher, this represents H for either the input or output. If the line drops lower, this represents L.

Some datasheets include schematics for various circuits that can be built around the particular component (see Figure 8-8). These diagrams are very useful in designing your circuitry and can be used as examples for building projects.

EN	1A	2A	FUNCTION
H	L	H	Turn right
H	H	L	Turn left
H	L	L	Fast motor stop
H	H	H	Fast motor stop
L	X	X	Fast motor stop

L = low, H = high, X = don't care

Figure 8-6   Function table

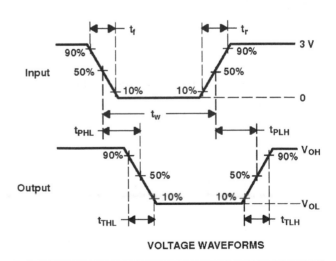

VOLTAGE WAVEFORMS

Figure 8-7   Timing bar

APPLICATION INFORMATION

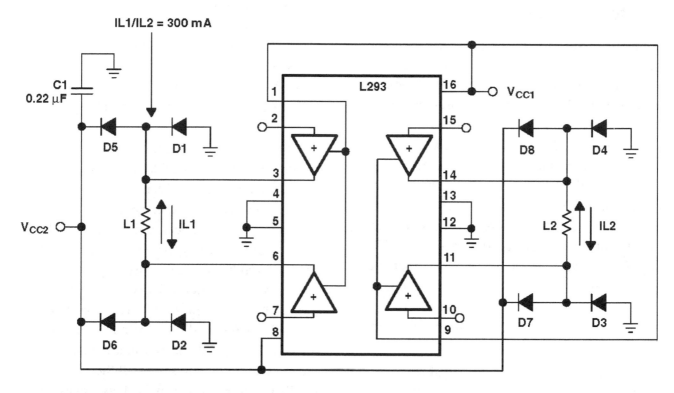

D1–D8 = SES5001

Figure 8-8 Schematic diagrams

Some parts can be static and sensitive to the way they are integrated into a circuit, in which case the datasheet will always provide layout considerations (see Figure 8-9). These features can range from noise-reduction techniques to dealing with thermal issues. For prototyping circuits, these considerations are generally not used. However, if you are creating a product design life cycle and implementing components into the real world or finished products, then taking the advice from this information will make for a trouble-free circuit. If guidelines are not followed, it becomes more difficult to diagnose problems and harder to fix them.

It is worth noting that some datasheets come with errors, just like any other technical document. Therefore, it's always best to make sure you have the very latest technical datasheet. Make sure you read the datasheet from

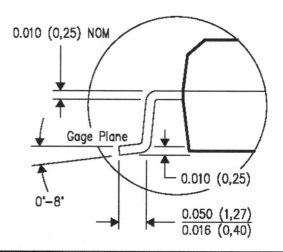

Figure 8-9 Mounting information

beginning to end to get a good understanding of all the features and characteristics; you must be absolutely certain that this is the part for your project. If in doubt, you must consult someone or contact the manufacturer.

Breadboards and Prototyping Boards

A breadboard is usually a rectangular plastic box with lots of little holes in it (see Figure 8-10). The holes are contacts into which you can insert electrical components or wires with ease. Breadboards are often used to put together a concept design of a circuit without having to solder any components; instead, you just poke the wires or the legs of the component into the holes, thus creating the contacts. The contacts are usually arranged in rows by connecting the metal contacts underneath the breadboard. The best thing about using a breadboard is that you can change the circuit design at any given point, so you can replace or rearrange components with ease without having to solder/unsolder any joints.

When you place components in a breadboard, not much happens unless you connect jumper wires up to create an electrical circuit. The wire used in electronics is made of copper surrounded by an outer plastic insulation, usually called hookup wire. Wire comes in all sorts of sizes (diameter), often referred to as the gauge. The standard measurement in the United States is AWG. It is always advisable to used solid wire rather than stranded wire because solid wire inserts into the breadboard much easier than does stranded wire. If you are lucky, your electronics shop will sell jumper wires, which are short lengths of wire with a single pin on each end (see Figure 8-11).

If you create your circuit design on a breadboard, you may decide that you want to make it permanent by soldering the components in place on a printed circuit board. To do this, you may have to get a universal printed circuit board, which in some ways is similar to a breadboard layout. A prototyping printed circuit board has rows of individual holes across it, much like all the pins on a breadboard (see Figure 8-12). Generally, all the components will go on the top of the board, and you solder underneath. When you solder all the wires, they usually go underneath as well. This makes for a much neater and cleaner board to work with and helps you avoid a lot of congestion if you are soldering a lot of components.

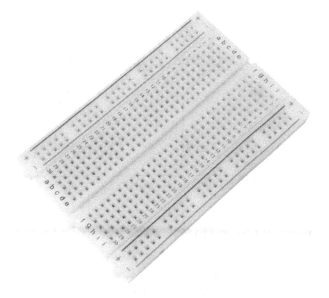

Figure 8-10   Breadboard

Figure 8-11   Jumper wires

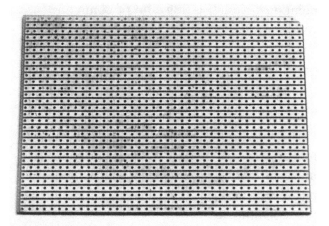

Figure 8-12 Prototyping board

Multimeter

A multimeter is a very useful device used to measure electricity, just like you would use a ruler to measure distance or a stopwatch to measure time. The best thing about a multimeter is that it also measures a lot of other things, such as voltage, current, resistance, and much more. A standard multimeter will have a large dial in the middle that lets you select what you want to measure (see Figure 8-13).

Most multimeters can measure voltage, current, and resistance; some multimeters

Figure 8-13 A typical multimeter

also have a continuity check that tests to see whether an electrical circuit is complete by creating a loud beeping sound when two items are electrically connected. This is very helpful in diagnosing problems with a circuit: you can trace the voltage around the circuit and find which part is incomplete or not functioning the way it should. Alternatively, you can use a multimeter to make sure that two items are *not* connected, just in case you don't want a certain part of your circuit to short—or you may want to test how good your soldering skills are by not accidentally soldering joints together.

There are some advanced multimeters, as well as expensive ones that have certain additional functions, such as the ability to measure transistors or capacitors. These functions are more suited for professional engineers who design and manufacture high-end products.

In order to understand how a multimeter works, it is important that you know what you are measuring, such as voltage, current, or resistance:

- Voltage is how hard the electricity is being pushed through a circuit; a higher voltage indicates that electricity is being pushed through the circuit really hard. Voltage uses the symbol V.

- Current is how much electricity is flowing through a circuit; a high current indicates that more electricity is flowing through the circuit. The symbol for current is A.

- Resistance is how difficult it is for electricity to flow through a circuit; a higher resistance indicates that it is more difficult for electricity to flow through the circuit. Resistance is measured in ohms, and the symbol for resistance is Ω.

It is also worth noting that the symbol being used for a unit of measurement may differ from the symbol for a variable equation. This is shown in Table 8-1.

TABLE 8-1	Electrical Symbols		
Variable	**Symbol**	**Unit**	**Symbol**
Voltage	V	Volts	V
Current	I	Ampere	A
Resistance	R	Ohm	Ω

A multimeter can be quite confusing to look at, with all the symbols—especially when you don't see words such as "voltage" and "current." To identify some of the features on a multimeter, Table 8-1 shows some of the commonly found symbols, which we can use to identify the unit of measurement. Most multimeters use the metric system:

- μ (micro) equals one millionth.
- m (milli) equals one thousandth.
- k (kilo) equals one thousand.
- M (mega) equals one million.

For example, when you see a 200k resistor, you know that the value is 200,000 ohms.

Some multimeters have an auto-ranging feature, and others require you to manually select the unit of measurement. If you need to manually select the range, it is usually best practice to elect a higher value than the one you are expecting to measure. For example, if you are going to measure a AA battery, you know that the voltage is roughly going to be 1.5V, which is standard for AA batteries. Therefore, you wouldn't select 200mV because the value would be too small. Instead, you would pick the next higher value, which is 2V. All the other range values would be too high and would be deemed unnecessary and result in a loss of accurate data.

When you purchase a new multimeter, you will notice that it comes with black and red wire probes (see Figure 8-14). One end of the wire comes with a banana jack, which plugs directly into your multimeter, and the other end is a probe tip used to measure the voltage

point on a circuit. Using a bit of common sense, you know that red always represents positive and black negative. Although you only get two probes, there can be up to four or five places on the multimeter to plug the banana plugs into, depending on what you want to measure.

Most multimeters will also have a fuse to protect them from too much current. When too much current is encountered, the fuse simply burns out and stops the circuit to avoid damage to the multimeter.

Measuring Voltage

Follow these steps to measure the voltage in a circuit:

1. Plug the black and red probes into the colored sockets on the multimeter. For most multimeters, the black probe plugs into the COM socket and the red probe plugs into the socket labeled V.

2. Choose the voltage range you are going to measure using the dial in the middle of the

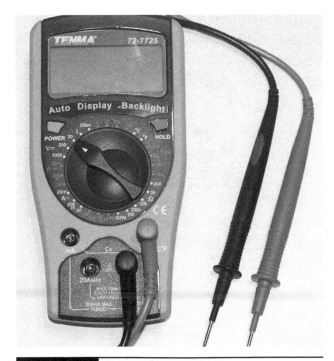

Figure 8-14   Test probes

multimeter. If you know roughly what the voltage is from the source, such as battery or a microcontroller, you can easily determine the range.

3. When touching the probe tips to your circuit, you must do this in parallel. If you want to measure the voltage across an LED, you would touch the black probe where the negative power source is and the red probe would be the positive source (in other words, you measure the points on either side of the LED).

4. If your voltage reading is not showing up or is displaying an error message on the screen, you may want to adjust the settings to a higher voltage reading and then try to read the voltage again.

Measuring Current

Follow these steps to measure the current in a circuit:

1. Plug the black and red probes into the colored sockets on the multimeter. For most multimeters, the black probe plugs into the COM socket and the red probe plugs into the socket labeled V.

2. Choose the appropriate current settings on the multimeter using the dial in the middle. Remember to check whether you are using alternating current or direct current. If you are using a battery, this will be direct current. If your multimeter is not auto-ranging, you will need to guess the current range (you can always change it afterward).

3. When measuring current, you should connect the probes in a series circuit. The best way to think of this is to pretend the multimeter is a part of your circuit, and just like any other component, you connect it up with the positive and negative points.

Measuring Resistance

Follow these steps to measure the resistance of a component in a circuit:

1. Plug the black and red probes into the colored sockets on the multimeter. For most multimeters, the black probe plugs into the COM socket and the red probe plugs into the socket labeled V.

2. Choose the value of the measurement using the dial. If you know the value of the component you are measuring, this can be used as an indicator to select the measurement on the multimeter to the nearest decimal point.

3. When measuring voltage, first turn off the power from the power source; otherwise, you may get an incorrect reading when measuring across two points. When measuring the resistance, always pick two points. Remember that the resistance is the same in both directions, so it does not make any difference which way you connect the black and red probes in the circuit.

Soldering

Soldering is an essential skill to learn in the world of electronics. Although you can get by just using a prototyping board, you may still need to solder headers onto the board or make some small modifications to a component.

Solder refers to the alloy that typically comes on a wire spool or tube, as shown in Figure 8-15, and it's this solder that we use to fuse components together on a printed circuit board. Solder usually comes in two types: leaded and lead-free solder. When solder was first used, it was generally made of an alloy of both lead and tin. Since then, it has become known that lead can be quite harmful to humans when exposed to it in large amounts. Lead was used in solder

Figure 8-15   Solder spool

because it has a great low melting point, and it created really good solder joints for very reliable circuit boards. Unfortunately, in the EU, leaded solder is not RoHS (Restriction of Hazardous Substances) compliant, and this restricts the use of leaded solder in electrical equipment. This is why lead-free solder is commonly used. Lead-free solder is usually made using other metals such as silver and copper. Lead-free solder does come with its own downfalls; it has a much higher melting point because of the increased tin content. As such, it requires a high-powered soldering iron.

Many types of lead-free solder contain a flux core that helps give it the same quality as leaded solder. Flux is a chemical agent that aids in the flowing of solder and creates a much better contact when finished.

Many tools aid in soldering, but none is more important than the soldering iron. Soldering irons can come in a variety of factors, ranging from the basic soldering iron to complex soldering stations, but at the end of the day, they all serve the same function and purpose. Usually, a good place to start is to buy a station composed of a soldering iron with a digital or analog controller and a stand. These stations are becoming more common now and can be very inexpensive to purchase at your local store.

Over time, your soldering tip can start to oxidize and will turn black. This is bad because the soldering iron will not cling to the solder; therefore, you cannot solder your component. This is more common with lead-free solder. This is where a simple sponge comes to the rescue. Every so often, you should try to clean the tip by wiping all the excess off with a soft sponge. For even better results, you can use a brass wire sponge (see Figure 8-16).

Apart from a soldering iron and solder, many great accessories can aid in the process of soldering. Solder wick is a vital tool for mopping up if you have made a bit of mess, or it can be used in desoldering. Solder wick is made up of thin copper braiding, and just like any PCB, it will soak up the solder, erasing any excess drops.

Using tip tinner is another method of cleaning your tip (see Figure 8-17). It is composed of a mild acid that helps remove any unwanted residue left on your soldering tip. This helps prevent oxidization when the tip is not in use.

As previously mentioned, some lead-free solder comes with a core flux; however, sometimes it is not enough, and extra flux may

Figure 8-16   Brass sponge

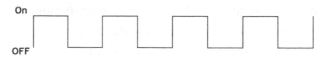

Figure 8-18 | Digital signal

Figure 8-17 | Tip tinner

be required. Flux pens are used to create a better bond between difficult components and the PCB.

Analog vs. Digital

Analog and digital signals are used to transmit an array of information, usually conveyed through electrical signals. The main difference between the two signal types is that analog signals are transmitted in pulses of varying amplitude whereas digital signals are transmitted in a binary format (ones and zeros), where each bit represents a distinct amplitude.

Analog refers to circuits where voltage or current varies at a continuous rate over a period of time. Electronic signals represent information by changing their voltage or current over time. The signal takes any value in a given range, and each signal value can represent different kinds of information. Any change that takes place in the signal has a significant impact on the overall result of information.

Something important to take into account is that analog signals can create noise, which is classified as a disturbance or variation. This can be caused by thermal vibrations. Because any slight variation in the signal can affect the outcome, this noise can have a greater effect, especially over long distances, because the signal degrades over distance.

When you are designing a system, analog circuits are much harder and complex to use, and require more skill compared to digital systems. This is primarily why digital systems have become more common, not to mention the fact that they are much cheaper to manufacture.

Digital systems are much easier to understand because they do not use a continuous range like analog. Therefore, any noise or slight variation in the signal does not impact the result of a digital signal. Digital systems generally have only two states, represented by two different voltages: usually 0 would be equal to ground and +V would be equal to 1 (see Figure 8-18).

The main advantage of using a digital system is that, compared to analog, the signal does not degrade over time, and it can be quite easily replicated without any loss. The main disadvantage compared to analog systems is that digital circuits consume much more power, and when circuits consume more power, that generally means more heat, which in turn increases the complexity of designing a circuit.

Understanding I2C and SPI

This chapter explains the differences between using I2C and SPI devices. When creating projects, it can be an important factor when selecting the hardware, as both protocols come with advantages and disadvantages.

I2C

Inter-integrated circuit (or I2C as everybody knows it) is a computer bus originally invented by Philips. It is used to attach low-speed peripherals to an embedded system or digital electronic circuit. This means that we can have multiple peripherals connected to one single device using only the unique I2C addressing system and four wires. I2C use a two-wire bidirectional line, serial data (SDA) and serial clock (SCL), with pull-up resistors. The I2C bus has a seven-bit address system to identify each device (0x00) and usually operates in two modes: master and slave. Each transmission between two devices always starts with a START command and ends with STOP. In between, both devices will send an ACK, stating they are ready to receive.

As previously mentioned, there is a serial clock line that holds the line after receiving or sending a byte, indicating that it is not yet ready to process more data. The master communicating with the slave device may not finish the transmission of the current bit until the clock line goes high.

Whenever communicating with a slave device, the master always checks the bus for a **STOP** or **START** message and does not start transmitting until it has received one of those two messages on the bus.

To display I2C devices on the BeagleBone Black, you can use the i2cdetect utility already installed:

```
# i2cdetect -y -r 1
     0  1  2  3  4  5  6  7  8  9  a  b  c  d  e  f
00:          -- -- -- -- -- -- -- -- -- -- -- -- --
10: -- -- -- -- -- -- -- -- -- -- -- -- -- -- -- --
20: -- -- -- -- -- -- -- -- -- -- -- -- -- -- -- --
30: -- -- -- -- -- -- -- -- -- -- -- -- -- -- -- --
40: -- -- -- -- -- -- -- -- -- -- -- -- -- -- -- --
50: -- -- -- -- UU UU UU UU -- -- -- -- -- -- -- --
60: -- -- -- -- -- -- -- -- -- -- -- -- -- -- -- --
70: 70 -- -- -- -- -- -- --
```

You can connect up to 127 different I2C devices on the I2C bus, as long as each device has its own address.

SPI

The serial peripheral interface (SPI) is for synchronous serial data communication between devices operating in full duplex mode. SPI was designed by Motorola specifically for short-distance communication in embedded systems. Similar to the I2C bus, the SPI uses a master/slave setup, where the master initiates the data. SPI uses a four-wire bus containing three-, two-, and one-wire serial busses.

The SPI bus specifies four logic signals:

- **SCLK**   Serial clock
- **MOSI**   Master output, slave input
- **MISO**   Master input, slave output
- **SS**   Slave select

Some peripherals with the SPI bus have been designed to be connected in a daisy-chain configuration, thus allowing multiple slaves to be connected to a single master. This daisy-chain communication is accomplished using shift registering by clock pulses; this is considered an advanced feature of SPI and is not available on some devices.

The BeagleBone Black has up to five serial ports available for use. Each serial port contains RX (receive) and TX (transmit) pins for two-way communication.

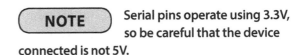 Serial pins operate using 3.3V, so be careful that the device connected is not 5V.

Summary

This chapter has been through some of the important and essential tips that you will need to create your own Evil Genius projects. Some of the tools you will use can be a little tricky at first, but with some practice, you will soon be a pro, and these tools will soon be an everyday item for your projects.

Suppliers and Components

ALL THE PARTS USED IN THIS BOOK are easily available for purchase from a number of stores through the Internet. However, sometimes it can become quite difficult to find exactly what you are looking for—especially in your own country. Finding suppliers in your own country will reduce the cost of large shipments, thus offering you a more affordable solution.

Suppliers

When searching for your components on the Internet, you will probably come across many suppliers all offering something different in terms of product selection. While writing this book, I came to use four main suppliers for all my components. These suppliers offer fantastic service as well as excellent pricing and product choice. When I'm looking for components and don't have to worry about the high minimum-order quantity, I use CPC in the United Kingdom or MCM in the United States.

CPC in the United Kingdom is a leading electrical distributer, offering over 100,000 different products from over 1,200 of the leading manufacturers. The vast array of components offers customers a one-stop shop for all their prototyping needs while providing free delivery in the United Kingdom for online orders (see www.cpc.co.uk).

MCM in the United States is a broadline distributor of electronic components and equipment for the consumer market. MCM is committed to providing all its customers top-quality service by providing over 40,000 products from over 600 vendors (see www.mcmelectronics. com).

Although both CPC and MCM source some great low-cost components, there are also some smaller companies offering niche products, such as Adafruit Industries and SparkFun. These companies often specialize in the maker market and also manufacture their own products.

SparkFun, based in the United States, is an online retail store that sells bits and pieces to make your electronic projects. In addition to its products, SparkFun offers classes and online tutorials designed to help educate individuals in embedded electronics (see www.sparkfun.com).

Adafruit Industries was founded in 2005 by MIT engineer Limor Fried. Adafruit was designed to create the best place online for learning about electronics and making the best designed products in the market. Adafruit designs and develops its own products in house and sells these products on its website, with lots of tutorials to get you started (see www.adafruit.com).

The sections that follow list components by type, along with some sources and order codes to make it easier for you to purchase your components.

Components

The component tables list appendix codes for each component that is used in that particular project. Table A-1 lists all the parts and offers some sources as to where they can be purchased.

TABLE A-1	BeagleBone Black and Modules	
Code	Description	Source
M1	BeagleBone Black	CPC: SC13491 MCM: 83-16241 Adafruit: 1876 SparkFun: DEV-12857
M2	Ultrasonic sensor	Adafruit: 980 SparkFun: SEN-08503
M3	Adafruit GPS module	Adafruit: 746 Pimoroni
M4	Prototyping shield	CPC: SC12957 MCM: 83-14776 Adafruit: 572
M5	I2C four-digit 7-segment display	Adafruit: 878 Pimoroni
M6	16×2 LCD screen	CPC: SC12098 MCM: 28-17797 Adafruit: 823 SparkFun: LCD-10862
M7	8×8 LED matrix backpack	CPC: SC12981 MCM: 25-6050 Adafruit: 959
M8	Electret microphone	Adafruit: 1713 Pimoroni
M9	Ultrasonic module	CPC: SN36696 MCM: 28-17967

Resistors

Resistors are low-cost components—almost less than one cent each. Often, suppliers will sell them in packs of 50 or 100. There are common resistors that get used a lot, such as 220R, 270R, 1K, and 10K values, so it can be very useful to keep a stock of these values.

After a while, you might find yourself buying a lot of resistors, and in some cases, it is better to buy them in kits, which stock the most popular values used in everyday electronics.

Here are some companies that sell resistor kits:

- MCM: 80-6665
- CPC: SC13464
- SparkFun: COM-10969

Table A-2 lists resistors, and Table A-3 lists capacitors.

TABLE A-2	Resistors	
Code	Description	Source
R1	220Ω 1/4W resistor	CPC: RE03795 MCM: 66-220
R2	110Ω 1/4W resistor	CPC: RE03728 MCM: 970-100
R3	4.6KΩ 1/4W resistor	CPC: RE03811 MCM: 34-4.7k
R4	10KΩ potentiometer	CPC: RE06540
R5	10KΩ 1/4W resistor	CPC: RE03815 MCM: 34-10k
R6	330Ω 1/4W resistor	CPC: RE03797 MCM: 34-330
R7	Photocell	CPC: RE04698 MCM: MC12-0100
R8	1KΩ 1/4W resistor	CPC: RE03803 MCM: 34-1k
R9	3.3KΩ 1/4W resistor	CPC: RE03809 MCM: 66-3.3k
R10	470Ω 1/4W resistor	CPC: RE03799 MCM: 34-470

TABLE A-3	Capacitors	
Code	Description	Source
C1	100nF	CPC: CA05514 Adafruit: 753 SparkFun: COM-08375
C2	100uF	CPC: CA07510 MCM: 31-5275

Semiconductors

The projects in this book use a lot of LEDs, so sometimes it is worth looking around for a variety pack of 5mm or 10mm LEDs rather than buying all the size and color combinations separately. Table A-4 lists semiconductors.

TABLE A-4   Semiconductors

Code	Description	Sources
S1	5mm red LED	CPC: SC11574 MCM: 25-5666 Adafruit: 297 SparkFun: COM-09590
S2	High power LED 1W	CPC: SC11783 MCM: 25-4770 Adafruit: 518 SparkFun: BOB-09656
S3	8mm RGB LED	SparkFun: COM-09264 Adafruit: 159
S4	5mm yellow LED	CPC: SC11577 SparkFun: COM-09594 MCM: NTE3146
S5	5mm green LED	CPC: SC11573 Adafruit: 298 SparkFun: COM-09650
S6	TMP36 temperature sensor	CPC: SC10437 Adafruit: 165 SparkFun: SEN-10988
S7	L293DNE motor driver	CPC: SC10241 Adafruit: 807 SparkFun: COM-00315
S8	1W audio amp	CPC: SCTDA2822M NEWARK: 89K1049
S9	Motion PIR sensor	CPC: SN36764 MCM: 28-17976 Adafruit: 189 SparkFun: SEN-08630
S10	TIP120 transistor	CPC: SC11000 MCM: 115-TIP120 Adafruit: 976
S11	1N4004 diode	CPC: SC07335 MCM: 107-1N4004
S12	LED array bar	CPC: SC12043

Hardware and Miscellaneous

Most of the products in this section are readily available in the majority of maker/hobbyist stores and are usually very inexpensive.

TABLE A-5   Hardware and Miscellaneous

Code	Description	Description
H1	Jumper wires	CPC: PC01769 MCM: 21-12365 Adafruit: 758 SparkFun: PRT-08431
H2	Solderless breadboard	CPC: PC01770 MCM: 28-17932 Adafruit: 64 SparkFun: PRT-09567
H3	Miniature push to make switch	CPC: SW04107 MCM: 26-2555 Adafruit: 1119 SparkFun: COM-00097
H5	Piezo transducer	CPC: LS02988 MCM: 28-11800 SparkFun: SEN-10293
H6	5V servo motor	CPC: MC02062 MCM: 28-17452 Adafruit: 169 SparkFun: ROB-09065
H7	6V DC motor	CPC: MC02075 MCM: 28-17657
H8	Robotic chassis	MCM: 28-17654 SparkFun: ROB-10825
H9	Wireless keyboard	CPC: CS23874 MCM: 83-14947 Adafruit: 922
H10	12V water pump	CPC: MC02068 MCM: 28-17461 Adafruit: 1150 SparkFun: ROB-10455
H11	USB Thunder Missile Launcher	Dream Cheeky Amazon eBay
H12	8W miniature speaker	CPC: LS00533 MCM: 55-4532 SparkFun: COM-09151

(continued on next page)

TABLE A-5	Hardware and Miscellaneous	
Code	Description	Description
H13	5V buzzer	CPC: LS03781 MCM: 287-1354 Adafruit: 160 SparkFun: COM-07950
H14	4×3 matrix keypad	CPC: SW04507 Adafruit: 419 SparkFun: COM-08653
H15	Solenoid/mag lock	CPC: SW04774, SR04745 MCM: 28-17455
H16	Logitech C270 webcam	CPC: CS20156 MCM: 83-12543
H17	USB audio card	CPC: CS21652 MCM: 83-15107

BeagleBone Black GPIO Pinout

TABLES **B-1** AND **B-2** SHOW the GPIO pinouts for the BeagleBone Black. This is a very useful guide when working with the BeagleBone Black and helps to determine the logistics of cable routing when prototyping.

TABLE B-1	GPIO Pinout for Pin Header P9		
GND	1	2	GND
3.3V	3	4	3.3V
5V RAW	5	6	5V RAW
5V	7	8	5V
	9	10	
Serial4 RX	11	12	GPIO P9_12
Serial4 TX	13	14	PWM1A
GPIO P9_15	15	16	PWM1B
GPIO P9_17	17	18	GPIO P9_18
I2C2_SCL	19	20	I2C2_SDA
Serial2 TX	21	22	Serial2 RX
GPIO P9_23	23	24	Serial1 TX
GPIO P9_25	25	26	Serial1 RX
GPIO P9_27	27	28	GPIO P9_28
GPIO P9_29	29	30	GPIO P9_30
GPIO P9_31	31	32	VDD_ADC
AIN4	33	34	GND_ADC
AIN6	35	36	AIN5
AIN2	37	38	AIN3
AIN0	39	40	AIN1
GPIO P9_41	41	42	GPIO P9_42
GND	43	44	GND
GND	45	46	GND
GND	47	48	GND

TABLE B-2	GPIO Pinout for Pin Header P8		
GND	1	2	GND
GPIO P8_3	3	4	GPIO P8_4
GPIO P8_5	5	6	GPIO P8_6
GPIO P8_7	7	8	GPIO P8_8
GPIO P8_9	9	10	GPIO P8_10
GPIO P8_11	11	12	GPIO P8_12
PWM2B	13	14	GPIO P8_14
GPIO P8_15	15	16	GPIO P8_16
GPIO P8_17	17	18	GPIO P8_18
GPIO P8_19	19	20	GPIO P8_20
GPIO P8_21	21	22	GPIO P8_22
GPIO P8_23	23	24	GPIO P8_24
GPIO P8_25	25	26	GPIO P8_26
GPIO P8_27	27	28	GPIO P8_28
GPIO P8_29	29	30	GPIO P8_30
GPIO P8_31	31	32	GPIO P8_32
GPIO P8_33	33	34	GPIO P8_34
GPIO P8_35	35	36	GPIO P8_36
Serial5 TX	37	38	Serial5 RX
GPIO P8_39	39	40	GPIO P8_40
GPIO P8_41	41	42	GPIO P8_42
GPIO P8_43	43	44	GPIO P8_44

Index

CPSIA information can be obtained
at www.ICGtesting.com
Printed in the USA
LVHW061700120520
655459LV00008B/624

9 780071 839280